HANDBOOK OF GEOGRAPHICAL NICKNAMES

HAROLD S. SHARP

THE SCARECROW PRESS, INC.
METUCHEN, N.J., & LONDON
1980

Library of Congress Cataloging in Publication Data

Sharp, Harold S
   Handbook of geographical nicknames.

    1. Names, Geographical. 2. Nicknames. I. Title.
G105.S5   910'.321   79-26860
ISBN 0-8108-1280-0

Dedicated to my good friends:
Betty and Paul Claymiller
Clara and Sam Cutright
Peach and Ted Medearis
Margaret and Ting Newman,
and to the memory of Doug Manke

FOREWORD

The word "nickname" is defined as "a name added to or substituted for the proper name of a person, place, etc., as in ridicule or familiarity." The purpose of this handbook is to indicate the nicknames of various cities, rivers, mountains, seas, deserts, capes, and the like.

Nicknames have come to be applied to geographical areas for a number of reasons. Some are the result of location, as in the case of Wilson's Promontory, known as "Australia's Farthest South," and Hammerfest, Norway, called "The Farthest North." An appellation may be based on mythology, as "The Abode of Apollo and the Muses," by which the mountain of Helicon, in Greece, is known. Hamlet, Shakespeare's play, led to Helsingör (Elsinore), Denmark, becoming known as "The Town of Hamlet." Lew Pollack and Sidney Clare's popular song of the twenties, "Chicago," led to State Street in that city being called "That Great Street." Joaquin Miller, in his poem "Columbus," speaks of "The Gray Azores" and "The Gates of Hercules." The latter appellation, used by the ancients, refers to two peaked rocks flanking the Strait of Gibraltar.

Rich agricultural areas generated several "garden" nicknames, including "The Garden of England" (Hereford), "The Garden of Scotland" (Morayshire), "The Garden of Spain" (Andalusia), and "The Garden of the Argentine" (Tucumán). Izalco, an active volcano, is called "The Lighthouse of El Salvador" because it creates a glow visible from the Pacific Ocean. Cities associated with religious history have come to be known as "The Holy City," examples of which are Jerusalem, Rome, and Benares. The Erie Canal has been called "Clinton's Big Ditch" and "Clinton's Ditch" because of the work of De Witt Clinton, the American lawyer and statesman who saw its construction through to completion.

Some areas bear a geographic resemblance to others. Thus we have Saint John's, Newfoundland, called "The Gibraltar of North America," both Aden and Corregidor nicknamed "The Gibraltar of the East," and Pearl Harbor referred to as "The Gibraltar of the Pacific." Saint Andrews, Scotland, the site of a famous golf course, became "The Golfer's Paradise." The goods produced locally led to Santos, Brazil, being spoken of as "The City of Coffee," Tours, France, becoming known as "The City of Silk," and Armagnac, France, being called "The Home of Armagnac Brandy." A historical event is sometimes responsible for the nickname of the place where it occurred, as Mainz, Germany, being known as "The Birth-place of Printing" and Runnymede, England, becoming "The Birth-place of the Magna Charta" and "The Birthplace of English Justice." Tulle, France, is mentioned as "The City of the Black Death" be-cause of the great plague epidemic that centered there in the four-teenth century.

This book does not indicate nicknames of American cities, states, or counties. These are adequately covered by Joseph Nathan Kane and Gerard L. Alexander's Nicknames and Sobriquets of U.S. Cities, States and Counties, third edition, published by Scarecrow Press in 1979. Ancient names and names in foreign languages are not given, not being nicknames; they may be determined from stand-ard encyclopedias. The reference departments of most public and university libraries also include assortments of geographical dic-tionaries, many of which contain this information.

All items, both the generally accepted names, shown as main entries, together with their applicable nicknames, appear in one alphabetical list, with the nicknames, in lower case, also being listed alphabetically and cross-referenced to the main entries, which appear in CAPITAL LETTERS. Each main entry is followed by a brief notation giving its location and other identification, with the applicable nickname(s) then shown. For example:

Main entry:

SAINT VINCENT, CAPE. Cape sit.
    at S. W. point of Portugal, about
    118 mi. S. of Lisbon

    The Sacred Promontory
    The Westernmost Point of
        Europe

Cross references:

CAPE. See: SAINT VINCENT, CAPE
Europe, The Westernmost Point of.
    See: SAINT VINCENT, CAPE
Point of Europe, The Westernmost.
    See: SAINT VINCENT, CAPE
Promontory, The Sacred. See:
    SAINT VINCENT, CAPE
Sacred Promontory, The. See:
    SAINT VINCENT, CAPE
VINCENT. See: SAINT VINCENT,
    CAPE
Westernmost Point of Europe, The.
    See: SAINT VINCENT, CAPE

As indicated in the example above, cross-referencing is

shown from all significant words. When a nickname covers an area so general that no name can be applied to it, the nickname (in lower case) becomes the main entry, thus:

Main entry:

Hell on Wheels.   Appellation of
    newly-formed American towns
    during the mid-19th century

Cross-reference:

Wheels. See: Hell on Wheels

Gathering material for a book of this nature is not a do-it-yourself project.   Many people gave valuable assistance while it was in preparation.   Particular thanks are due the reference librarians at the Fort Wayne and Allen County Library, The Indiana/Purdue Library (Fort Wayne), the Indianapolis Public Library, and the Indiana University Library (Bloomington).   The compiler wishes to express his gratitude to all these willing helpers at this time.

Harold S. Sharp
Fort Wayne, Indiana
1980

-A-

ABAJO. See: VUELTA ABAJO

ABBAZIA, CROATIA. Village of Croatia, Yugoslavia, 5 mi. W. of Rieka

The Nice of the Adriatic

ABBEY. See: RAMSAY ABBEY

Abbey, Westminster. See: COLLEGIATE CHURCH OF ST. PETER

Abbeys, The Croesus of English. See: RAMSAY ABBEY

ABERDEEN, SCOTLAND. City; capital of Aberdeenshire, about 130 mi. N. E. of Edinburgh

The Granite City
The Silver City by the Sea

Abode of Apollo and the Muses, The. See: HELICON

Abode of Snow, The. See: HIMALAYA, THE

Abode of the Gods, The. See: IDA MOUNTAINS

ABRAHAM. See: PLAINS OF ABRAHAM

ABYSSINIA. See: ETHIOPIA (ABYSSINIA)

Achilles, The Residence of. See: PHTHIA, GREECE

Achilles' Heel of England, The. See: IRELAND (EIRE)

Achilles of Rivers, The. See: COLUMBIA RIVER

ACHIN, N. SUMATRA. Native sultanate, now a government, Netherland Indies

The Montenegro of Sumatra

Acre. See: Hell's Half Acre

ADEN, S. W. ARABIA. Seaport of Aden Protectorate on Gulf of Aden, 110 mi. E. of Bab el Mandeb Strait

The Gibraltar of the East

Adonis, The City of. See: JUBAYL, LEBANON

Adriatic, The Bride of the. See: VENICE, ITALY

Adriatic, The Mistress of the. See: VENICE, ITALY

Adriatic, The Nice of the. See: ABBAZIA, CROATIA

Adriatic, The Queen of the. See: VENICE, ITALY

Aeolian Islands, The. See: LIPARI ISLANDS

AFRICA. Second largest continent on the globe; approx. 11,530,000 sq. mi.

Brightest Africa
The Dark Continent
Darkest Africa
The Land of the Fetish

AFRICA. See also: EQUATORIAL AFRICA; SOUTH AFRICA

Africa, Europe's. See: BALKAN MOUNTAINS

Africa, The Head of. See: GOOD HOPE, CAPE OF

Africa, The Hero of. See: CUBA

Africa, The Horn of. See: SOMALILAND, AFRICA

Africa, The Montenegro of. See: ETHIOPIA (ABYSSINIA)

Africa, The Most English Town in South. See: GRAHAMSTOWN, SOUTH AFRICA

Africa, The Westernmost Point of. See: ALMADIES, CAPE

Africa's Farthest West. See: VERDE, CAPE

Agamemnon's Capital. See: MYCENAE, GREECE

AGAR-TOWN ENGLAND. Section of metropolitan borough of St. Pancras, London

The English Connemara

Agony, The Grotto of. See: GETHSEMANE, PALESTINE

AGRA, INDIA. City of India, sit. on right bank of Jumna

R., 110 mi. S. E. of Delhi

The Rome of Hindustan

Agriculture, The Home of Early Chinese. See: SHANSI, CHINA

Air, The West Point of the. See: RANDOLPH FIELD, TEXAS

Aircraft Carriers, The Stationary Island. See: MARIANA ISLANDS

ALAMO, THE. Fort located in San Antonio, Texas; besieged by Mexicans under Santa Anna February 23--March 6, 1836

The Thermopylae of America

ALBANIA. Nation of Europe sit. in W. part of Balkan Peninsula

The Eagle's Country

Albert the Bear, The City of. See: DESSAU, GERMANY

Albion. See: ENGLAND; GREAT BRITAIN; New Albion

ALCANTARA, SPAIN. Town of Spain sit. in former region of Estremadura 7 mi. E. of Portuguese border

The Bridge

ALCATRAZ ISLAND. Island sit. in San Francisco Bay, California; former military and federal prison

America's Devil's Island
America's First Escape-
   Proof Prison
Bird Island

The Island
The Island of Pelicans
The Isle of No Return
Pelican Island
The Rock
The Tarnished Jewel

ALENTEJO, PORTUGAL. Province of S. E. Portugal sit. on
Atlantic Ocean

The Granary of Portugal

ALEXANDRIA, EGYPT.
Egyptian seaport sit. between
Mediterranean Sea and Lake
Mareotis W. of Abukir Bay

The Delta City
The Mother of Books
The Pearl of the Mediter-
ranean

All-British Colony, The City of
the. See: LLOYDMINSTER,
CANADA

ALLAHABAD, INDIA. City of
India located at confluence
of Jumna and Ganges Rivers,
350 mi. S. E. of New Delhi

The Holy City

Alley. See: Tin Pan Alley;
Tornado Alley

Alluvial Gold, The City of.
See: BENDIGO, AUSTRALIA

ALMADIES, CAPE. Extreme
tip of peninsula forming Cape
Vert, Senegal, West Africa

The Westernmost Point of
Africa

Alps, The Gap in the. See:
BRENNER PASS

ALPS, SWISS. Portion of
mountain system of S. central Europe extending from
Mediterranean coast between

France and Italy into Switzerland

The Helvetian Mountains

Alps, The Tiger of the. See:
MATTERHORN, THE

Altar Stones, The Home of the.
See: STONEHENGE, ENG-
LAND

ALTDORF, SWITZERLAND.
Commune of Uri Canton,
Switzerland, near S. E. tip
of Lake Lucerne, 20 mi.
S. E. of Lucerne

The Home of William Tell

AMAZON RIVER. River of S.
America flowing from Peru-
vian Andes to Atlantic Ocean;
about 3700 mi. long

The King of Waters
The Mediterranean of Brazil

America. See: Little Amer-
ica; Main Street of America,
The; Newmarket of America,
The; UNITED STATES OF
AMERICA

America, The Archaeological
Capital of South. See: CUZ-
CO, PERU

America, The Athens of South.
See: BOGOTA, COLOMBIA

America, The Backbone of
North. See: ROCKY
MOUNTAINS

America, The Eden of. See:
AQUIDNECK ISLAND

America, The Gibraltar of.
See: QUEBEC, CANADA

America, The Gibraltar of
North. See: SAINT JOHN'S,
NEWFOUNDLAND

America, The Heroin Crossroads of South. See: PAR-
AGUAY

America, The Little Switzerland
of. See: REST ISLANDS,
MINN.

America, The Rhine of. See:
HUDSON RIVER

America, The Rooftop of Eastern. See: GREAT SMOKY
MOUNTAINS

America, The Switzerland of
Central. See: PANAMA

America, The Thermopylae of.
See: ALAMO, THE

America, The Venice of. See:
RECIFE, BRAZIL

America, The Wonderland of.
See: YELLOWSTONE NA-
TIONAL PARK

American Business, The Grand
Canyon of. See: BROAD-
WAY, NEW YORK CITY

American Continent, The Backbone of the. See: ROCKY
MOUNTAINS

American Desert. See: Great
American Desert, The

American Liberty, The Cradle
of. See: FANEUIL HALL;
INDEPENDENCE HALL

American Nile, The. See:
SAINT JOHN'S RIVER

AMERICAN RIVER. River of
N. central California, about
30 mi. long, flowing into
Sacramento R.; site of gold
strike of 1848

The River of Gold

American Trade, The Sixteenth
Century Emporium of South.
See: PORTO BELLO, PAN-
AMA

America's Bermuda. See:
BLOCK ISLAND

America's Devil's Island. See:
ALCATRAZ ISLAND

America's Farthest South. See:
SABLE, CAPE

America's First Escape-Proof
Prison. See: ALCATRAZ
ISLAND

America's Great Winter Garden. See: IMPERIAL VAL-
LEY

AMHERST, NOVA SCOTIA.
Town of Cumberland Coun-
ty, Nova Scotia, 5 mi. E.
of N.E. end of Chignecto Bay.

The Inland Gateway to Nova
Scotia

AMIENS, FRANCE. City of
Somme dept., Northern
France, sit. on Somme R.,
72 mi. N. of Paris

Little Venice
The Venice of France

Ammonites, The Chief City of
the. See: PHILADELPHIA,
PALESTINE

Amon, The City of. See:
THEBES, EGYPT

AMRITSAR, INDIA. Capital
city of Punjab, Republic of
India, 32 mi. S.E. of La-
hore

The City of the Golden Temple

The City of the Pool of Immortality

AMYCLAE, GREECE. Ancient city of Laconia, S. E. Peloponnesus, S. Greece, 3 mi. S. of Sparta

The Silent City

Ancient Violet-Crowned Athens. See: ATHENS, GREECE

Ancient World. See: Seven Wonders of the Ancient World, The

Ancient World, The Paris of the. See: CORINTH, GREECE

Ancients, The Mountain of Ten Thousand. See: WAN-SHOU-SHAN

ANDALUSIA, SPAIN. Region of Spain reaching across southern part of the country

The Garden and Granary of Spain
The Garden of Spain
The Gold Purse of Spain
The Granary of Spain

ANDERSONVILLE PRISON. Confederate Civil War military prison maintained at Andersonville, Georgia

Hell on Earth

ANDES MOUNTAINS. Mountain system of South America extending along west coast from Tierra del Fuego to Panama; about 4500 mi.

The Treasury of Peru

ANDREWS. See: SAINT ANDREWS

ANGERS, FRANCE. Capital

of the dept. of Maine-et-Loire, France, located on Maine R. 48 mi. E. N. E. of Nantes

Black Angers
The Black City

ANGKOR, CAMBODIA. Ruined ancient city of Cambodia located about 1 mi. N. of Angkor Wat

The Grand Capital

ANGLESEY, ENGLAND. Island on N. W. coast of Wales; separated from mainland by Menai Strait

The Place of the Druids

ANTARCTICA. Portion of earth's surface surrounding the South Pole lying mainly within the Antarctic Circle

The Bottom of the World
The Ice Island
The Land of Frozen Time

Anti-Fascist Protection Wall, The. See: BERLIN WALL

Antibes, The Pearl of the. See: CUBA

Antilles, The Pearl of the. See: CUBA

ANTIOCH, ASIA MINOR. Ancient city of Pisidia, Asia sit. near Yalvac, Turkey in Asia, 80 mi. W. N. W. of Konya

Antioch the Beautiful
The Crown of the East
The Mother of Christian Missions
The Queen of the East

ANTWERP, BELGIUM. City of Belgium sit. on right

bank of Schelde R. 23 mi.
N. E. of Brussels

The Diamond Capital of the
World

APENNINES. Mountain range
in central Italy extending
full length of the peninsula
from near Savona to Reggio
di Calabria

The Purple Apennines

Apollo, The City of. See:
GRYNEION, ASIA MINOR

Apollo and the Muses, The
Abode of. See: HELICON

APOSTLE ISLANDS. Group
of 20 islands in S. W. Lake
Superior off coast of Wisconsin

The Twelve Apostles

Apostle of God, The City of
the. See: MEDINA, SAU-
DI ARABIA

APPIAN WAY. First paved
Roman road leading from
Rome to Brundisium

The Queen of Roads

AQUIDNECK ISLAND. Island
in Narragansett Bay, an inlet of the Atlantic Ocean in
S. E. Rhode I.

The Eden of America
The Isle of Peace

AQUILEIA, ITALY. Town of
Friuli province, Italy, at
head of Adriatic Sea, 22 mi.
N. W. of Trieste

The Second Rome

ARABIA. Peninsula of S. W.

Asia, about 1400 mi. long
and 1250 mi. wide

The Holy Land

ARABIAN DESERT. Arid re-
gion of Egypt bounded by
Mediterranean Sea, Gulf of
Suez and Nile R.

The Eastern Desert

Arabian Klondike, The. See:
LIBYA (LIBIA)

Arabian Nights, The City of the.
See: BAGHDAD, IRAQ

ARAL, SEA OF. Inland sea sit.
between Kazakh S. S. R. and
Uzbek S. S. R. , Soviet Russia

The Island Sea

ARAN ISLANDS. Group of 3
small islands off W. Coast
of Ireland at entrance to
Galway Bay

Aran of the Saints

ARANJUEZ, SPAIN. Commune
of Madrid province, central
Spain 26 mi. S. S. E. of Madrid

The Metropolis of Flora

Archaeological Capital of South
America, The. See: CUZ-
CO, PERU

ARCHIPELAGO. See: MALAY
ARCHIPELAGO

Archipelago. See: Ritchie's
Archipelago

Archipelago, The Queen of the
Eastern. See: JAVA

ARENDAL, NORWAY. Seaport
of Norway on the Skagerrak;

seat of Aust-Adger county,
S. Norway

Little Venice

Argentine, The Garden of the.
See: TUCUMAN

Aristotle, The Town of. See:
STAGIRA; GREECE

ARLBERG, AUSTRIA. Alpine
valley, pass and tunnel in
Tirol, W. Austria

The Home of the Arlberg
Skiing Technique

ARLES, FRANCE. Medieval
kingdom formed from union
of kingdoms of Cisjurane
Burgundy and Transjurane
Burgundy

The Kingdom of Burgundy

ARMAGH, IRELAND. County
of S. Northern Ireland sit.
on S. shore of Lough Neagh

The Orchard of Ireland

ARMAGNAC, FRANCE.
Small territory in old
province of Gascony,
S. W. France

The Home of Armagnac
Brandy

Armory of Germany, The.
See: SUHL, GERMANY

Asia, The Gate of. See:
KAZAN, U.S.S.R.

Asia, The Gateway to Central.
See: KHYBER PASS

Assizes, The Home of the
Bloody. See: DORCHESTER,
ENGLAND

ASWAN, EGYPT. City of

S. E. Upper Egypt sit. on
right bank of Nile R. oppo-
sit Elephantine Island

The Egyptian Resort City

Athens, The German. See:
WEIMAR, GERMANY

Athens, The Granary of. See:
KERCH, U.S.S.R.

ATHENS, GREECE. Capital
city of Greece sit. in S. E.
portion, about 5 mi. E. of
Piraeus on Saronic Gulf

Ancient Violet-Crowned Ath-
ens
The City of the Violet Crown
The Eye of Greece
The Religious City
The Violet-Crowned City

Athens, The Modern. See:
EDINBURGH, SCOTLAND

Athens, The Mohammedan.
See: BAGHDAD, IRAQ

Athens, The Northern. See:
EDINBURGH, SCOTLAND

Athens of Ireland, The. See:
BELFAST, IRELAND; CORK,
IRELAND

Athens of South America, The.
See: BOGOTA, COLOMBIA

Athens of Switzerland, The.
See: ZURICH, SWITZER-
LAND

Athens of the North, The.
See: COPENHAGEN, DEN-
MARK; EDINBURGH, SCOT-
LAND

Athens of the West, The. See:
CORDOVA, SPAIN

ATHLONE, IRELAND. Urban
district of N. Central Ire-

land sit. on Shannon R.

The Heart of Ireland

ATHOS, MOUNT. Mountain
at E. end of Acte peninsula,
N. E. Greece, 6670 feet high

The Holy Hill
The Holy Mountain

Atlantic, The Fisherman's Paradise of the North. See:
BLOCK ISLAND

Atlantic, The Pearl of the.
See: MADEIRA

ATLANTIC OCEAN. Body of
salt water lying between the
E. coast of America and the
W. Coasts of Europe and
Africa

The Herring Pond
The Sea of Darkness
The Western Ocean

ATLANTIS. Mythical island
of antiquity in Western
Ocean, supposedly located
near Pillars of Hercules;
said to have sunk beneath
the ocean surface

The Sunken Continent
The Sunken Island

Auld. See also: Old; Owld

Auld Reekie. See: EDIN-
BURGH, SCOTLAND

AUSTRALIA. Island continent
lying between Pacific and
Indian Ocean S. of New
Guinea

Down Under
The Island Continent
New Holland
The Old Country

Australia, The Montpellier of.

See: BRISBANE, AUS-
TRALIA

Australia's Farthest South. See:
WILSON'S PROMONTORY

Australia's Farthest West. See:
STEEP POINT

Australia's Sunshine Coast.
See: NOOSA HEADS

AUSTRIA-HUNGARY. Former
monarchy in Central Europe,
including the empire of Austria, kingdom of Hungary and
various crownlands; dissolved
in 1918

The China of Europe
The Dual Monarchy

AVERNUS, LAKE. Lake sit.
near Pozzuoli, Italy, 8 mi.
W. of Naples; mentioned in
Vergil's Aeneid

Hell's Entrance
The Legendary Mouth of
Hell

AVIV. See: TEL AVIV, IS-
RAEL

AVON. See: STRATFORD-
UPON-AVON, ENGLAND

AXMINSTER, ENGLAND. Town
of Devonshire, England, sit.
on Axe R., 23 mi. E. N. E.
of Exeter

The Carpet City

Ayacucho, The Peace of. See:
LA PAZ, BOLIVIA

AZORES. Group of 9 volcanic
islands and several islets
in N. Atlantic Ocean

The Gray Azores

AZOV, SEA OF. Inland sea

in S. European Soviet Russia
connected with Black Sea by
Strait of Kerch

The Fish Sea

Azure Coast, The.   See: RIV-
IERA, THE

-B-

BAALBEK, LEBANON.   Village
of E. Lebanon Republic, 35
mi. N. of Damascus; very
important in ancient times

The City of the Sun
The Solar City

BAB EL MANDEB.   Strait
between Arabia and Africa
by which Red Sea connected
with Gulf of Aden

The Gate of Tears

BABYLON.   Ancient city sit.
on Euphrates R. 55 mi.
S. of Baghdad

The Exactress of Gold

Babylon, The Modern.   See:
LONDON, ENGLAND

BABYLONIA.   Ancient country
of Mesopotamia lying be-
tween Tigris and Euphrates
rivers; roughly coextensive
with present-day Baghdad
Province, Iraq

The Land of the Chaldees

Backbone, The Devil's.   See:
NATCHEZ TRACE

Backbone of North America,
The.   See: ROCKY MOUN-
TAINS

Backbone of the American

Continent, The.   See: ROCKY
MOUNTAINS

Backbone of the Confederacy,
The.   See: MISSISSIPPI
RIVER

Backward, The River That
Runs.   See: CHICAGO
RIVER

Bad Lands, The.   Rugged region of fantastically shaped
rock masses and hills vir-
tually without vegetation sit.
in western part of Dakotas
and Western Nebraska

BADAJOZ, SPAIN.   Spanish
province sit. on Portuguese
border

The Land of Health

BADEN, AUSTRIA.   Commune,
lower Austria province, Austria, 14 mi. S. S. W. of Vienna

The Eden of Germany

BADEN-BADEN, GERMANY.
City of Karlsruhe district,
Baden, Germany, 18 mi.
S. S. W. of Karlsruhe

The European Saratoga

BAGHDAD, IRAQ.   City of
Iraq sit. on both sides of
Tigris R.

The City of Peace
The City of the Arabian
Nights
The Mohammedan Athens

Bailiwick, Wat Tyler's.   See:
DARTFORD, ENGLAND

BAKU, U. S. S. R.   City of
Azerbaidzhan Republic,
U. S. S. R. , sit. on W. Coast
of Caspian Sea

The Portsmouth of the
Steppes

BALKAN MOUNTAINS. Range
of mountains extending E.
and W. across Central Bul-
garia from Yugoslav border
to Black Sea

Europe's Africa

BALKH, BACTRIA. Ancient
city of Bactria located in
central Asia

The Mother of Cities

Baltic, The Eye of the. See:
GOTLAND

BALTIC SEA. Sea of N.
Europe; arm of Atlantic
Ocean

The Mediterranean of the
North

BALTISTAN, INDIA. Part of
Ladakh frontier district, N.
central Kashmir State, N.
India

Little Tibet

Banal Frontier, The. Part
of former military frontier
of Austrian Empire

BANBURY, ENGLAND. Town
of Oxfordshire, England,
23 mi. N. of Oxford

The Home of Banbury Cakes
The Town of Banbury Tarts

BANGKOK, THAILAND. City
of Thailand on Chao R.
about 20 mi. above its
mouth

The Rice Bowl of the World
The Venice of the East

BANK. See: GRAND BANK

Banker to Half the World, The.
See: HONG KONG, CHINA

BARATARIA BAY. Inlet of
Gulf of Mexico on boundary
between Jefferson and Pla-
quimines parishes, S. E.
Louisiana

The Pirate's Home

BARBADOS. British island
and colony in Lesser An-
tilles, W. Indies, E. of
central Windward Islands

Little England

Barbary Coast, The. Former
red light district of San Francisco, California

Barbary Coast of the East, The.
See: SANDS STREET, BROOK-
LYN, N. Y.

Barbary Pirates' Stronghold,
The. See: TRIPOLI, LIBYA

Bard, The Home of the. See:
STRATFORD-UPON-AVON,
ENGLAND

BARISAL, PAKISTAN. Town
of Bakarganj district, Dakka
division, E. Bengal, Pakistan, 130 mi. E. of Calcutta

The Town of the Barisal
Guns

Basin, The Red. See: SZECH-
WAN, CHINA

BASS STRAIT. Strait separating Australia from Tasmania;
80 to 150 mi. wide

The Straits

Bastille, The English. See:
COLDBATH FIELDS

BASTOGNE, BELGIUM. Town

of E. Luxembourg province,
S. E. Belgium, 43 mi. S.
of Liége

Germany's Waterloo
Hitler's Waterloo

BATAVIA, NETHERLAND IN-
DIES. Residency, N. W.
Java province, Netherland
Indies

The Queen of the East

BATH, ENGLAND. City and
county borough, Somerset-
shire, N. W. England, sit.
on Avon R. , 12 mi. E. S. E.
of Bristol

The Queen of Spas

BAVARIA. Former German
State of S. Germany now
sit. in American zone following World War II

The Beer-Producing State

BAY. See: BARATARIA
BAY; CHESAPEAKE BAY;
CROMER BAY; FUNDY,
BAY OF; INUTIL BAY

Bay, Useless. See: INUTIL
BAY

Bay of Tides, The. See:
FUNDY, BAY OF

Bayonet, The Birthplace of
the. See: BAYONNE,
FRANCE

BAYONNE, FRANCE. City
of Bassess-Pyrénées dept. ,
S. W. France, 55 mi. W. N. W.
of Pau

The Birthplace of the Bayonet

BEACH. See: WAIKIKI
BEACH

Beach East, Miami. See:
TEL AVIV, ISRAEL

Beantown. See: FRIJOLES

Bear, The City of Albert the.
See: DESSAU, GERMANY

Bear, The Northern. See:
RUSSIA

Beautiful, The. See: NAPLES,
ITALY

Beautiful, Antioch the. See:
ANTIOCH, ASIA MINOR

Beautiful Blue Danube, The.
See: DANUBE RIVER

Beautiful Daughter of Rome,
The. See: FLORENCE,
ITALY

Bed, The Home of the Great.
See: WARE, ENGLAND

Bedford Level. See: Fen
Country, The

Bedlam. Corruption of "Bethlehem, " hospital of St. Mary's
of Bethlehem, London, England, founded in 1247 as a
priory, afterwards used as
a lunatic asylum

BEDLOE'S ISLAND. Island and
national monument sit. in
New York Harbor; site of
Berthold's statue "Liberty
Enlightening the World" ("Sta-
tue of Liberty")

Liberty Island

Beer, The Home of Bock. See:
EINBECK, WEST GERMANY

Beer, The Home of Pilsener.
See: PLZEN (PILSEN)

Beer Hall Putsch, The Town of
the. See: MUNICH, BAVARIA

Beer-Producing State, The.
See: BAVARIA

Bees, The Empire of. See:
HYBLA, SICILY

Beethoven, The Birthplace of.
See: BONN, GERMANY

BELFAST, IRELAND. County
borough and seaport of E.
Northern Ireland

The Athens of Ireland

BELGIUM. Constitutional
monarchy of N.W. Europe
sit. on N. Sea

The Cockpit of Europe
The Garden of Europe

Belgium, The Birmingham of.
See: LIEGE, BELGIUM

Belgium, The Manchester of.
See: GHENT, BELGIUM

Bell Rock, The. See: INCH-
CAPE ROCK

BELLO. See: PORTO BELLO,
PANAMA

Bells, The City of. See:
STRASBOURG, FRANCE

Belt. See: Bible Belt, The;
Black Belt, The

BENARES, INDIA. City on
Ganges R. , India, 400 mi.
W. N. W. of Calcutta

The Holy City

Benares of the South, The.
See: CONJEEVERAM,
INDIA

BENDIGO, AUSTRALIA. City
in Victoria, Australia, 80
mi. N.N.W. of Melbourne

The City of Alluvial Gold

Benelux. Collective designation for Belgium, the Netherlands and Luxembourg

BERGE. See: HÖRSEL
BERGE, GERMANY

BERING SEA. Part of N. Pacific Ocean lying between
Aleutian Islands on the S.
and the Bering Strait with
which it is connected to the
Arctic Ocean on the N.

The Sea of Kamchatka

BERLIN, GERMANY. City of
Berlin province, Prussia,
Germany, sit. on Spree R.
163 mi. S. E. of Hamburg

The City of Intelligence

BERLIN WALL. Wall erected
by Russians in 1961 separating East from West Ger-
many

The Anti-Fascist Protection
Wall
The Wall of Infamy

Bermuda, America's. See:
BLOCK ISLAND

Bermuda Triangle, The. Area
of Atlantic Ocean where
many ships and aircraft
have mysteriously disap-
peared

Bernard, The Home of St.
See: CLAIRVAUX, FRANCE

BETHANY, PALESTINE. Small
village sit. on Mount of
Olives about 2 mi. E. of
Jerusalem on Jericho Road

The Home of Lazarus

BETHEL, PALESTINE. Ruined
ancient village 11 mi. N. of
Jerusalem

The House of God

BETHESDA. Pool or public
bath in Jerusalem at time
of Christ

The House of Mercy

Bethlehem. See: Bedlam

BETHLEHEM, PALESTINE.
Town of Palestine sit.
5 mi. S. of Jerusalem;
birthplace of Jesus Christ

The Home of David
The House of Bread
The House of Deity
The Site of the Nativity

Between Two Rivers, The
Country. See: MESO-
POTAMIA

BHAGALPUR, INDIA. Cap-
ital of district of same name,
Bihar state, India, on
Ganges R. 205 mi. N. N. W.
of Calcutta

Tiger City

Bible, The Birthplace of the
Douai. See: DOUAI,
FRANCE

Bible, The City of the Brothers'. See: KRALICKA,
CZECHOSLOVAKIA

Bible, The City of the Kralitz.
See: KRALICKA, CZECH-
OSLOVAKIA

Bible Belt, The. Rural area
of S. and S. W. United
States

Biblical City, The. See:
TYRE, LEBANON

Big Ditch, The. See: PANA-
MA CANAL

Big Ditch, Clinton's. See:
ERIE CANAL

Big Island, The. See: HA-
WAII

Big Muddy River. See: MIS-
SOURI RIVER

BIG SALMON RIVER. R. of
central Idaho, flowing from
S. Custer County to Snake
R. ; about 420 mi. long

The River of No Return

Big Smoke, The. See: LON-
DON, ENGLAND

Big Three Conference, The
Place of the. See: YALTA,
RUSSIA

BIJAPUR, INDIA. Town of
S. E. Bombay province, India, 240 mi. S. E. of Bombay

The Palmyra of the Deccan

BIMINI. Mythical island sought
by Ponce de Leon, Spanish
explorer

The Island of the Fountain
of Youth

Bird Island. See: ALCATRAZ
ISLAND

Birdcage Walk. Burial ground
of Newgate Prison, London,
England

Bird's Head, The. See: VO-
GELKOP

Birmingham of Belgium, The.
See: LIEGE, BELGIUM

Birmingham of Russia, The.

See: TULA, RUSSIA

Birthplace of Beethoven, The.
  See: BONN, GERMANY

Birthplace of Columbus, The.
  See: GENOA, ITALY

Birthplace of English Justice,
  The. See: RUNNYMEDE,
  ENGLAND

Birthplace of Liberty, The.
  See: INDEPENDENCE
  HALL

Birthplace of Printing, The.
  See: MAINZ, GERMANY

Birthplace of the Bayonet,
  The. See: BAYONNE,
  FRANCE

Birthplace of the Cooperative
  Movement, The. See:
  ROCHDALE, ENGLAND

Birthplace of the Douai Bible,
  The. See: DOUAI, FRANCE

Birthplace of the Magna Charta,
  The. See: RUNNYMEDE,
  ENGLAND

Birthplace of the Pistol, The.
  See: PISTOIA, ITALY

Black Angers. See: ANGERS,
  FRANCE

Black Belt, The. Strip of
  prairie land along coast of
  Alabama and Mississippi
  with black, clayey soil

Black City, The. See: AN-
  GERS, FRANCE

Black Country, The. Midland
  districts of S. Staffordshire
  and N. Warwickshire, England

Black Death, The City of

the. See: TULLE, FRANCE

Black Desert, The. See: KA-
  RA KUM

Black Elster, The. See:
  SCHWARZE ELSTER

Black Hills, The. Group of
  mountains in W. So. Dakota
  and N. E. Wyoming

Black Hole, The City of the.
  See: CALCUTTA, INDIA

Black Hole of Calcutta, The.
  Small chamber at Fort William, Calcutta, India, stronghold of East India Company

Black Mountain, The. See:
  MONTENEGRO

Black Republic, The. See:
  DOMINICAN REPUBLIC;
  HAITI

BLACK SEA. Inland sea lying
  between Eastern Europe and
  Asia Minor and connected
  with the Agean Sea

  The Hospitable Sea
  The Inhospitable Sea

Black Sea, The Queen of the.
  See: ODESSA, RUSSIA

Blacks, The Country of the.
  See: SUDAN, AFRICA

Blackwell's Island. See: WEL-
  FARE ISLAND

Blades, The City of Sword.
  See: TOLEDO, SPAIN

BELEKINGE, SWEDEN. Swed-
  ish province sit. in S. E. part
  of the country and adjoining
  the Black Sea

  The Garden of Sweden

Blessed, The Islands of the.
See: CANARY ISLANDS

Blighty. Expression used by
British soldiers in World
War I signifying "home"

BLOCK ISLAND. Island in
Atlantic Ocean sit. at E. entrance to Long Island; coextensive with town of New
Shoreham, R. I.

America's Bermuda
The Fisherman's Paradise
of the North Atlantic

Blood. See: Field of Blood,
The

Bloody Assizes, The Home of
the. See: DORCHESTER,
ENGLAND

Bloody Bridge, The. Bridge
over Walloomsac R. , Ben-
nington County, Vermont,
35 mi. from Troy, N. Y.

Bloody Nose Ridge. See:
UMURBROGOL

Bloody Pond, The. Pool near
Fort Edward, New York;
grave of soldiers killed in
French and Indian War

Bloody Tower, The. See:
TOWER OF LONDON

Blue Danube, The. See:
DANUBE RIVER

Blue Danube, The Beautiful.
See: DANUBE RIVER

Blue Mosque, The. See:
MASJID JABUD

Boar, The Land of the. See:
GERMANY

Bock Beer, The Home of.

See: EINBECK, WEST GER-
MANY

BOGOTA, COLOMBIA. Capital
city of Colombia in E. cor-
dillera of Andes Mountains

The Athens of South America

Bohemia, The Center of. See:
GREENWICH VILLAGE

Bohemia, The Paradise of.
See: LEITMERITZ, BO-
HEMIA

Bohemian Paradise, The. See:
LEITMERITZ, BOHEMIA

Bohemianism, The Hotbed of.
See: GREENWICH VILLAGE

BOLOGNA, ITALY. Commune
of Italy sit. at foot of Apennine Mts. , 51 mi. N. by E.
of Florence

The Fat

Bondage, The Land of. See:
EGYPT

BONN, GERMANY. City of
Köln govt. district, Rhine
Province, Prussia, Germany,
sit. on left bank of Rhine R.
16 mi. S. S. E. of Cologne

The Birthplace of Beethoven

Book of Kells, The Home of
the. See: KELLS, IRELAND

Books, The Mother of. See:
ALEXANDRIA, EGYPT

Boone's Trail. See: WILDER-
NESS ROAD, THE

Boot, The Toe of the Italian.
See: CALABRIA, ITALY

BOOTHIA PENINSULA, CAN-

ADA. Peninsula in Franklin District, Northwest Territories, Canada; northernmost point of North American mainland

The Home of the North Magnetic Pole

BORDEAUX, FRANCE. Seaport and capital of department of Gironde, France, sit. on W. bank of Garonne R. 60 mi. from Atlantic Ocean

The Home of Wines

Border, North of the. See: CANADA

Border, South of the. See: MEXICO

Boroughs. See: Five Boroughs, The

BOSCOBEL, ENGLAND. Locality of Shropshire, W. England, E. of Shrewsbury

The Place of the Royal Oak

BOSPORUS. Strait connecting the Black Sea and the Sea of Marmara separating Turkey in Asia from Turkey in Europe

The Ford of the Ox
The Golden Horn

BOSSENO, FRANCE. Group of Celtic monuments and Gallo-Roman ruins approx. $1\frac{1}{2}$ mi. from Carnac, village in dept. of Morbihan, France

Caesar's Camp

Boston, The Wall Street of. See: STATE STREET,

BOSTON, MASS.

Bottom of the World, The. See: ANTARCTICA; SOUTH POLE

"Bounty" Island, The. See: PITCAIRN ISLAND

Bowl. See: Dust Bowl, The; Rice Bowl, The

Bowl of the World, The Rice, See: BANGKOK, THAILAND

Brahma, The Son of. See: BRAHMAPUTRA RIVER

BRAHMAPUTRA RIVER. River in India rising as the Tsangpo or Matsang in S.W. Tibet

The Son of Brahma

Branch of Honolulu, The Long. See: WAIKIKI BEACH

Branch of Philadelphia, The Long. See: MAY, CAPE

Brandy, The Home of Armagnac. See: CONDOM, FRANCE

Brandy, The Town of. See: COGNAC, FRANCE

Brave, The Land of the Free and the Home of the. See: UNITED STATES OF AMERICA

BRAZIL. Federal republic of E. central S. America

Half a Continent

Brazil, The Mediterranean of. See: AMAZON RIVER

Bread, The House of. See: BETHLEHEM, PALESTINE

BRENNER PASS. Alpine pass between Innsbruk, Tirol-Voralberg province, Austria, and Bressonone, Venezia Tridentina, Italy

The Gap in the Alps

Brickyard, The. Automobile racetrack in Speedway, Indiana, 5 mi. W. of Indianapolis

Bride of the Adriatic, The. See: VENICE, ITALY

Bride of the Sea, The. See: VENICE, ITALY

BRIDGE. See: BROOKLYN BRIDGE; SYDNEY HARBOR BRIDGE

Bridge. See: Bloody Bridge, The

Bridge, The. See: ALCANTARA, SPAIN

Bridge, The Great. See: BROOKLYN BRIDGE

Bridge of Sighs, The. Bridge between Doge's Palace and prison known as Carceri Prigioni at Venice, Italy, spanning the Rio della Paglia, a canal. Also ramp between former New York Criminal Courts Bldg. and adjoining prison, called the Tombs, New York City

BRIE, FRANCE. Agricultural district of N. E. France, E. of Paris

The Home of Brie Cheese

Briel, The Netherlands. See: Cautionary Towns, The

Brightest Africa. See: AFRICA

BRISBANE, AUSTRALIA. City of Queensland, Australia, sit. on N. bank of Brisbane R. near its mouth

The Montpellier of Australia

BRITAIN. See: GREAT BRITAIN

Britain. See: Greater Britain

Britain, Little. See: BRITTANY, FRANCE

Britain, North. See: SCOTLAND

Britain of the Far East, The. See: JAPAN

Britannia's Lion. See: GIBRALTAR

British, The Pantheon of the. See: COLLEGIATE CHURCH OF ST. PETER

British Colony, The City of the All. See: LLOYDMINSTER, CANADA

British Empire, The Jewel of the. See: INDIA

BRITTANY, FRANCE. Peninsular region of N. W. France sit. on English Channel

Little Britain

BROADWAY, NEW YORK CITY. Famous street of New York City, N. Y.

The Grand Canyon of American Business
The Great White Way
The Main Stem

BROKEN HILL, AFRICA. Town of Luangway province, central N. Rhodesia, S. central

Africa, 70 mi. N. of Lusaka

The Home of the Rhodesian Man

Broken Treaty, The Stone of the. See: LIMERICK, IRELAND

BROOKLYN BRIDGE. Bridge spanning E. River, New York City and connecting Brooklyn and Manhattan

The Eighth Wonder of the World
The Great Bridge
The Moonshot of 1883

Brothers' Bible, The City of the. See: KRALICKA, CZECHOSLOVAKIA

BRUSSELS, BELGIUM. Capital and largest city of Belgium, sit. on Senne R. about 27 mi. S. of Antwerp

Little Paris
The Village of the Marsh

Brute's City. See: LONDON, ENGLAND

BUDA, HUNGARY. Town of Hungary, now part of Budapest, on right bank of Danube R., 135 mi. S. E. of Vienna

The Key of Christendom

Buddhism, The Rome of. See: LHASA (LASSA), TIBET

Buddhism, The Vatican of. See: MANDALAY, BURMA

BUENA. See: YERBA BUENA

Bughouse Square. See: WASH-INGTON SQUARE

BUKHARA, ASIA. Former emirate sit. around city of Bukhara, W. Asia, later state in Russian Central Asia

The Treasury of Sciences

Burgundy, The Kingdom of. See: ARLES, FRANCE

BURMA. Republic of S. E. Asia sit. on Bay of Bengal

The Nation at the Crossroads

Burma Road, The Starting Point of the. See: LASHIO, BURMA

BURSLEM, ENGLAND. Town of Staffordshire, England, 20 mi. N. N. W. of Stafford; famous for pottery manufactured there

The Mother of the Potteries

Business, The Grand Canyon of American. See: BROADWAY, NEW YORK CITY

BUSIRIS, EGYPT. Ancient city of Egypt sit. in Nile Delta, 30 mi. S. W. of Tanis

The City of Osiris

Busy Ditch, The. See: PANAMA CANAL

-C-

C. C. C. HIGHWAY, THE. Highway running from Cleveland to Cincinnati, Ohio

The Three C's Highway

Cactus on a Stone. See: MEX-

ICO CITY, MEXICO

Caesar's Camp. See: BOS-
SENO, FRANCE

CAIRO, EGYPT. Capital city
of United Arab Republic,
sit. on E. bank of Nile R.
115 mi. S. E. of Alexandria

The City of Victory
The Tent
The Victorious City

Cakes, The Home of Banbury.
See: BANBURY, ENGLAND

Cakes, The Land of. See:
SCOTLAND

Cal, New. See: NEW CAL-
EDONIA

CALABRIA, ITALY. Region
of Italy comprising southernmost part of Italian peninsula and consisting of provinces of Catanzaro, Co-
senza and Reggio di Cala-
bria

The Toe of the Italian Boot

Calcutta. See: Black Hole
of Calcutta, The

CALCUTTA, INDIA. City of
S. W. Bengal, N. E. Indian
Union, sit. on Hooghly R. ,
90 mi. from its mouth

The City of Palaces
The City of the Black Hole

CALEDONIA. See: NEW
CALEDONIA

Caledonia. See: SCOTLAND

Calico, The City of. See:
CALICUT, INDIA

CALICUT, INDIA. City of W.

Madras, S. Indian Union,
on Malabar Coast, 350 mi.
W. S. W. of Madras

The City of Calico

California, The Granary of.
See: SAN JOAQUIN VALLEY

CALIFORNIA, GULF OF. Arm
of Pacific Ocean separating
peninsula of Lower California from the rest of Mexico

The Sea of Cortes
The Vermillion Sea

Calm, The Land of the Morning. See: KOREA

CAMBODIA. Kingdom, former
French protectorate, in Fed-
eration of Indochina, sit. on
Gulf of Siam

Where Past and Present
Meet

CAMBRAI, FRANCE. Indus-
trial city of Nord dept. , N.
France, sit. on Schelde R.
34 mi. S. by E. of Lille

The City of Cambresine
The City of Cambric

Cambresine, The City of. See:
CAMBRAI, FRANCE

Cambric, The City of. See:
CAMBRAI, FRANCE

Camel with the Feet of Gold,
The. See: LIBYA (LIBIA)

Camp, Caesar's. See: BOS-
SENO, FRANCE

CAMP CENTURY, GREENLAND.
American military base of
Greenland

The City Under the Ice

CAMP DAVID, MARYLAND.
Retreat used by the Presidents of the United States

The President's Retreat

Can Island, Tin.   See: NIUAFOO

CANAAN, PALESTINE.   Area
of Palestine sit. between
Jordan R. and Mediterranean Sea

The Land of Promise
The Promised Land

CANADA.   Federated state
of North America; largest
member of Commonwealth
of Nations

The Fourteenth Colony
North of the Border

CANAL.   See: ERIE CANAL;
GRAND CANAL; PANAMA
CANAL; SUEZ CANAL

CANARY ISLANDS.   Island
group in Atlantic Ocean
off N. W. coast of Africa,
823 mi. S. W. of Spain

The Fortunate Isles
The Happy Islands
The Islands of Dogs
The Islands of the Blessed

CANNON MOUNTAIN.   Peak
in White Mountains, W.
New Hampshire, on W.
side of Franconia Notch

The Great Stone Face
The Old Man of the Mountain
The Profile Mountain

Canoe, The Home of the Peterborough.   See: PETERBOROUGH, CANADA

Canoe, The Home of the Rice
Lake.   See: PETERBOROUGH, CANADA

CANOSSA, ITALY.   Village of
ruins of Castle where Henry
IV, King of Germany, did
public penance (1077)

The Town of Humble Submission

CANTERBURY, ENGLAND.
City and county of Kent,
England, sit. on Stour R.,
54 mi. E. S. E. of London

The Town of the Kentishmen

Cantons.   See: Forest Cantons,
The Four

Cantons, The Lake of the Four.
See: LUCERNE, LAKE OF

Cantons, The Lake of the Four
Forest.   See: LUCERNE,
LAKE OF

CANYON.   See: WAIMEA CANYON

Canyon of American Business,
The Grand.   See: BROADWAY, NEW YORK CITY

Canyon of the Pacific, The
Grand.   See: WAIMEA
CANYON

CAP-HAITIEN, HAITI.   Seaport on N. coast of Haiti,
80 mi. N. N. E. of Port-au-Prince

Le Cap
Little Paris

CAPE.   See: ALMADIES,
CAPE; COS, CAPE; DIAMOND, CAPE; DOUCATO,
CAPE; FINISTERRE, CAPE;
GOOD HOPE, CAPE OF;
HORN, CAPE; HOWE, CAPE;
MAY, CAPE; SABLE, CAPE;

SAINT VINCENT, CAPE;
VERDE, CAPE

Cape, The. See: COD, CAPE;
GOOD HOPE, CAPE OF;
HORN, CAPE

Cape, The Stormy. See:
GOOD HOPE, CAPE OF

Capital, Agamemnon's. See:
MYCENAE, GREECE

Capital, The Eastern. See:
DACCA, BANGLADESH;
TOKYO, JAPAN

Capital, The Grand. See:
ANGKOR, CAMBODIA

Capital, The Great. See:
KHANBALIK, CHINA

Capital, The Northern. See:
PEKING, CHINA

Capital, The Sky High. See:
LA PAZ, BOLIVIA

Capital, The Southern. See:
NANKING, CHINA

Capital of South America, The
Archaeological. See: CUZ-
CO, PERU

Capital of the Negro Population of the United States,
The. See: HARLEM

Capital of the World, The
Diamond. See: ANTWERP,
BELGIUM

Capital of the World, The Oil.
See: DHAHRAN, SAUDI
ARABIA

Capitol, The. That part of
the Capitoline Hill at Rome,
Italy, occupied by the temple
of Jupiter Optimus

CAPITOL HILL. Site of

building in Washington, D. C.,
used by the Congress for
its sessions

The Hill

CAPRI, ITALY. Island at entrance of Bay of Naples, Naples province, Italy; site of
the Blue Grotto

The Isle of Capri

CARIBBEAN SEA. Arm of Atlantic Ocean connecting with
Gulf of Mexico through Yucatan Channel

The Sea of the New World
The Spanish Main

CARLOW, IRELAND. County
of S. E. Ireland, Leinster
Province

The Garden of Erin

CARNATIC, INDIA. Region and
old division between E. Ghats
and Coromandel coast, S.
India

The Country of the Kanarese

Carpet, Tangipahoa of the Crimson. See: TANGIPAHOA

Carpet City, The. See: AX-
MINSTER, ENGLAND

Carpetani, The Stronghold of.
See: TOLEDO, SPAIN

Carpets, The Home of Kidderminster. See: KIDDER-
MINSTER, ENGLAND

Carpets, The Home of Turkish.
See: USAK, TURKEY

Carriers, The Stationary Island
Aircraft. See: MARIANA
ISLANDS

CARTHAGE, NORTH AFRICA.
Ancient city and state near
modern Tunis on peninsula
in small bay of Mediterranean Sea

Dido's City
The New Town

Carthage of the North, The.
See: LUBECK, GERMANY

CASHEL, IRELAND. Town of
S. central Tipperary, S.
Ireland

The City of the Kings

CASHMERE. See: KASHMIR
(CASHMERE), INDIA

Caspian Gates, The. See:
DERBENT, RUSSIA IN
EUROPE

Castle of Maidens, The. See:
EDINBURGH, SCOTLAND

Castle Stronghold, The. See:
EL QASR, EGYPT

CASTLEBAR, IRELAND. Ur-
ban district of County Mayo,
N. W. Ireland

The Home of the Castlebar
Races

Cat, The Isle of the Tailless.
See: MAN, ISLE OF

CATANIA, SICILY. Seaport,
Episcopal see and capital
of Catania province, Sicily,
59 mi. S. S. W. of Messina

The Granary of Sicily

Cathay. See: CHINA

Cathedral City, The. See:
NORWICH, ENGLAND

Cattle, The Home of Guern-

sey. See: GUERNSEY

Causes, The City of Lost.
See: GENEVA, SWITZER-
LAND

Causeway. See: Giant's Causeway, The

Cautionary Towns, The. Collective appellation of Netherlands towns of Briel, Flushing, Rammekens and Walcher-
en

Caves, The Home of the. See:
NOTTINGHAM, ENGLAND

CAWNPORE, INDIA. City, Al-
lahabad division, S. central
United Provinces, N. Indian
Union, sit. on N. bank of
Ganges R. 245 mi. S. E. of
Delhi

The City of the Massacre

Celestial City, The. See:
PEKING, CHINA

Celestial Empire, The. See:
CHINA

Celestial Mountains, The. See:
TIEN-SHAN

Center, The World's Diamond.
See: KIMBERLEY, SOUTH
AFRICA

Center of Bohemia, The. See:
GREENWICH VILLAGE

Center of Protestant Europe,
The Intellectual. See:
GENEVA, SWITZERLAND

Center of the Soviet Riviera,
The. See: YALTA, RUSSIA

Center of the World, The. See:
LAOS

Central America, The Switzerland of. See: PANAMA

Central Asia, The Gateway to.
  See: KHYBER PASS

CENTURY. See: CAMP CEN-
  TURY, GREENLAND

Century Emporium of South
  American Trade, The Six-
  teenth. See: PORTO BEL-
  LO, PANAMA

CEPHALONIA. Island in Ionian
  Sea, off W. coast of Greece

  Hidden Island
  Lost Island

Cervin, Mont. See: MATTER-
  HORN, THE

CEYLON. See: Sri Lanka

CHABLIS, FRANCE. Com-
  mune, Yonne dept. , N. E.
  central France, 11 mi. E.
  of Auxerre

  The Home of White Wines

Chaldees, The Land of the.
  See: BABYLONIA

Chamber, The Queen's. See:
  SPITHEAD, ENGLAND

CHANNEL. See: ENGLISH
  CHANNEL

Channel, The. See: ENGLISH
  CHANNEL

CHAPULTEPEC, MEXICO.
  Rocky height surmounted
  by historic castle about
  3 mi. S. W. of Mexico City

  Grasshopper Hill

CHARTREUSE, LA GRANDE.
  Carthusian monastery located
  about 12 mi. N. of Grenoble,
  France

The Home of Chartreuse
  Liqueur

CHEB, CZECHOSLOVAKIA.
  City of W. Bohemia pro-
  vince, W. Czechoslovakia,
  50 mi. N. W. of Pizen

The City of Wallenstein's
  Death

CHEDDAR, ENGLAND. Town
  of Somersetshire, England,
  22 mi. S. W. of Bristol

The City of Cheese

Cheese, The City of. See:
  CHEDDAR, ENGLAND; EDAM,
  THE NETHERLANDS

Cheese, The Home of. See:
  LIMBURG, BELGIUM

Cheese, The Home of Brie.
  See: BRIE, FRANCE

Cheese, The Home of Parmesan. See: PARMA, ITALY

Cheese, The Town of. See:
  ROQUEFORT-SUR-SOULZON,
  FRANCE

CHEMNITZ, GERMANY. City
  of E. Germany, in Saxony,
  43 mi. S. E. of Leipzig

The Saxon Manchester

Cherokee Strip. Narrow strip
  of land along S. border of
  Kansas; opened to settlers
  September 16, 1893, and became part of Oklahoma Territory

Chersonese, The Golden. See:
  MALAY PENINSULA (MAL-
  AYSIA)

CHESAPEAKE BAY. Inlet of
  Atlantic Ocean between states
  of Virginia and Maryland;

200 mi. long

The Tobacco Coast

Chicago, The Wall Street of.
See: LA SALLE STREET,
CHICAGO, ILL.

Chicago of Siberia, The. See:
NOVOSIBIRSK, RUSSIA

CHICAGO RIVER. River in
Chicago, Ill.; normal di-
rection of flow reversed during construction of Chicago
sewage disposal system

The River That Runs Back-
ward

CHICHEN ITZA. Ancient
ruined city of Yucatan
sit. 18 mi. S. W. of Val-
ladolid

The Mouth of the Wells of
Itza

CHICKAHOMINY RIVER. R.
in Virginia about 90 mi.
long; merges with James
R. 8 mi. W. of Williams-
burg

The River of Coarse Pounded Corn

Chief City of the Ammonites,
The. See: PHILADELPHIA,
PALESTINE

Chief of the Cinque Ports, The.
See: DOVER, ENGLAND

Children's Village, The. See:
PUSHKIN, RUSSIA

CHILE. Republic of S. W.
South America sit. between
Andes Mts. and Pacific
Ocean

The Shoestring Republic

CHILLON. Castle in Vaud,
W. Switzerland, at end of
Lake Geneva; place of im-
prisonment of François de
Bonnivard (1530-36)

The Home of the Prisoner

CHINA. Republic sit. at E.
end of central Asia

Cathay
The Celestial Empire
The Flowery Kingdom
The Flowery Land
The Hungry Dragon
The Middle Flowery Kingdom
The Middle Kingdom
The Sleeping Giant

China, The Birmingham of.
See: FATSHAN, CHINA

China, The Girdle of. See:
YANGTZE KIANG RIVER

China, The Great Commercial
Highway of Southeast. See:
SI-KIANG

China, The Great Wall of. See:
GREAT WALL, THE

China, The Home of Dresden.
See: DRESDEN, GERMANY

China, The Home of the Republic of. See: TAIWAN
(FORMOSA)

China, The Pittsburgh of. See:
HANKOW, CHINA

China, The Venice of. See:
WUHSIEN, CHINA

China of Europe, The. See:
AUSTRIA-HUNGARY

China's Sacred Mountain. See:
TAI-SHAN

China's Sorrow. See: HWANG
HO

Chinese Agriculture, The Home
of Early. See: SHANSI,
CHINA

Chinese Civilization, The Cradle of. See: WEI RIVER
VALLEY

Chinese Wall, The. See:
GREAT WALL, THE

Chinnereth, Sea of. See:
GALILEE, SEA OF

CHISHOLM TRAIL. Cattle
trail extending from near
San Antonio, Texas, to
Abilene, Kansas

John's Trail

Christendom, The Key of.
See: BUDA, HUNGARY

Christian Missions, The
Mother of. See: AN-
TIOCH, ASIA MINOR

Christmas Trees. See: Mile
of Christmas Trees, The

Chu Hsi, The Home of. See:
LUNGKI, CHINA

CHURCH. See: GOWRIE
CHURCH; Kirk; OLD
SOUTH CHURCH; SAINT
GILES'S CHURCH

Church, The States of the.
See: Papal States, The

CHURCH OF ST. PETER.
See: COLLEGIATE
CHURCH OF ST. PETER

Church on the Dunes. See:
DUNKIRK, FRANCE

Churches, A City of. See:
TORONTO, CANADA

CHURCHILL DOWNS, LOU-

ISVILLE, KY. Race track
sit. in Louisville, Ky.;
scene of annual Kentucky
Derby

The Home of the Derby

Cibola. See: Seven Cities of
Cibola, The

Cinque Ports, The Chief of the.
See: DOVER, ENGLAND

CIRCLE. See: COLUMBUS
CIRCLE, NEW YORK CITY

Circuit of the Northern Seas,
The. See: HOKKAIDO,
JAPAN

Citadel of the Orient, The.
See: SINGAPORE ISLAND

Cities. See: Han Cities, The;
League of Rhine Cities, The;
Seven Cities, The; Seven
Cities of Cibola, The; Six
Cities, The; Three Cities,
The

Cities, The Mother of. See:
BALKH, BACTRIA; KIEV,
U. S. S. R.

Cities, The Queen of. See:
ROME, ITALY

Cities of the Plain, The. See:
Gomorrah; Sodom

CITY. See: MEXICO CITY,
MEXICO; VATICAN CITY

City, The. See: LONDON,
ENGLAND

City, The Biblical. See:
TYRE, LEBANON

City, The Black. See: AN-
GERS, FRANCE

City, Brute's. See: LONDON

ENGLAND

City, The Carpet. See: AX-
MINSTER, ENGLAND

City, The Cathedral. See:
NORWICH, ENGLAND

City, The Celestial. See:
PEKING, CHINA

City, The Coronation. See:
VLADIMIR, U. S. S. R.

City, The Delta. See: AL-
EXANDRIA, EGYPT

City, Dido's. See: CARTH-
AGE, NORTH AFRICA

City, The Drisheen. See:
CORK, IRELAND

City, The Egyptian Resort.
See: ASWAN, EGYPT

City, The Enchanted. See:
PETRA, ARABIA

City, The Eternal. See:
ROME, ITALY

City, The Ever-Loyal. See:
OXFORD, ENGLAND

City, The Fair. See: PERTH,
SCOTLAND

City, The Forbidden. See:
Forbidden City, The;
LHASA (LASSA), TIBET;
PEKING, CHINA

City, The Golden. See:
JOHANNESBURG, SOUTH
AFRICA

City, The Granite. See:
ABERDEEN, SCOTLAND

City, Hill. See: GIBEON,
PALESTINE

City, The Hill. See: ROME,

ITALY

City, The Holy. See: ALLA-
HABAD, INDIA; BENARES,
INDIA; CUZCO, PERU; DA-
MASCUS, SYRIA; JERUSA-
LEM, PALESTINE; KIEV,
U. S. S. R.; MOSCOW, RUS-
SIA; ROME, ITALY

City, The Hundred-Gated. See:
THEBES, EGYPT

City, The Imperial. See:
ROME, ITALY

City, The Island. See: MON-
TREAL, CANADA

City, The Knuckle-Dusting.
See: LIVERPOOL, ENG-
LAND

City, The Leonine. See: Leonine City, The

City, The Lion. See: LEM-
BERG (LVOV), U. S. S. R.;
SINGAPORE, ASIA

City, The Museum. See: NOV-
GOROD, RUSSIA

City, The Nameless. See:
ROME, ITALY

City, The New. See: NAPLES,
ITALY

City, The Petrified. See:
ISHMONIE, EGYPT

City, The Prophet's. See:
MEDINA, SAUDI ARABIA

City, The Religious. See:
ATHENS, GREECE

City, The Sacred. See: JE-
RUSALEM, PALESTINE

City, The Seven-Hilled. See:
ROME, ITALY

City, The Silent. See: AMY-
CLAE, GREECE; VENICE,
ITALY

City, The Solar. See: BAAL-
BEK, LEBANON

City, The Sunken. See: PORT
ROYAL, JAMAICA

City, Tiger. See: BHAGAL-
PUR, INDIA

City, The Victorious. See:
CAIRO, EGYPT

City, The Violet-Crowned.
See: ATHENS, GREECE

City, The Violin. See: CRE-
MONA, ITALY

City by the Sea, The Silver.
See: ABERDEEN, SCOT-
LAND

City of a Hundred Towers,
The. See: PAVIA, ITALY

City of a Million Dreams, The.
See: VIENNA, AUSTRIA

City of Adonis, The. See:
JUBAYL, LEBANON

City of Albert the Bear, The.
See: DESSAU, GERMANY

City of Alluvial Gold, The.
See: BENDIGO, AUS-
TRALIA

City of Amon, The. See:
THEBES, EGYPT

City of Apollo, The. See:
GRYNEION, ASIA MINOR

City of Bells, The. See:
STRASBOURG, FRANCE

City of Calico, The. See:
CALICUT, INDIA

City of Cambresine, The. See:
CAMBRAI, FRANCE

City of Cambric, The. See:
CAMBRAI, FRANCE

City of Cheese, The. See:
CHEDDAR, ENGLAND;
EDAM, THE NETHERLANDS

City of Churches, A. See:
TORONTO, CANADA

City of Coffee, The. See:
SANTOS, BRAZIL

City of Dagon, The. See:
GAZA, PALESTINE

City of David, The. See:
JERUSALEM, PALESTINE;
TEMPLE HILL; ZION, PAL-
ESTINE

City of Demeter, The. See:
ELEUSIS, GREECE

City of Deportations, The.
See: LILLE, FRANCE

City of Diamonds, The. See:
GOLCONDA, INDIA

City of Herod, The Once-Royal.
See: TIBERIAS, PALESTINE

City of Homes, A. See: TOR-
ONTO, CANADA

City of Husain, The. See:
KERBELA, IRAQ

City of Intelligence, The. See:
BERLIN, GERMANY

City of Islam, The Holy. See:
MECCA, SAUDI ARABIA;
MEDINA, SAUDI ARABIA

City of Jagannath, The. See:
PURI, INDIA

City of Jouy Print, The. See:

JOUY-EN-JOSAS, FRANCE

City of Lace, The. See:
HONITON, ENGLAND

City of Light, The. See:
PARIS, FRANCE

City of Lilies, The. See:
FLORENCE, ITALY

City of Lost Causes, The.
See: GENEVA, SWITZER-
LAND

City of Masts, The. See:
LONDON, ENGLAND

City of Matches, The. See:
JÖNKÖPING, SWEDEN

City of Mosques, The. See:
DAMASCUS, SYRIA

City of Nergal, The. See:
CUTHAH, BABYLONIA

City of Osiris, The. See:
BUSIRIS, EGYPT

City of Palaces, The. See:
CALCUTTA, INDIA; EDIN-
BURGH, SCOTLAND; GEN-
OA, ITALY; LENINGRAD,
U. S. S. R.; PARIS, FRANCE;
ROME, ITALY

City of Peace, The. See:
BAGHDAD, IRAQ; GENEVA,
SWITZERLAND; JERUSA-
LEM, PALESTINE

City of Perspectives, The.
See: LENINGRAD, U. S. S. R.

City of Poseidon, The. See:
HELICE, GREECE

City of Pyramids, The. See:
GIZA, EGYPT

City of Pyrrhic Victory, The.
See: HERACLEA, ITALY

City of Queensland, The Garden. See: TOOWOOMBA,
AUSTRALIA

City of Ra, The. See: HEL-
IOPOLIS, EGYPT

City of Refuge, The. See:
RAMOTH GILEAD, PALES-
TINE

City of Roses, The. See:
LUCKNOW, INDIA

City of St. George, The. See:
LOD, PALESTINE

City of Seven Hills, The. See:
ROME, ITALY

City of Silk, The. See: TOURS,
FRANCE

City of Silver, The. See:
SUCRE, BOLIVIA

City of Smoke, The. See:
LONDON, ENGLAND

City of Snow, The. See: LEN-
INGRAD, U. S. S. R.

City of Sword Blades, The.
See: TOLEDO, SPAIN

City of the All-British Colony,
The. See: LLOYDMINSTER,
CANADA

City of the Ammonites, The
Chief. See: PHILADELPHIA,
PALESTINE

City of the Apostle of God, The.
See: MEDINA, SAUDI ARA-
BIA

City of the Arabian Nights, The.
See: BAGHDAD, IRAQ

City of the Black Death, The.
See: TULLE, FRANCE

City of the Black Hole, The.

See: CALCUTTA, INDIA

City of the Brothers' Bible,
The. See: KRALICKA,
CZECHOSLOVAKIA

City of the Dead, The. See:
VERACRUZ, MEXICO

City of the Golden Temple,
The. See: AMRITSAR,
INDIA

City of the Gordian Knot,
The. See: GORDIUM,
ASIA MINOR

City of the Great Khan, The.
See: KHANBALIK, CHINA;
PEKING, CHINA

City of the Great King, The.
See: JERUSALEM, PAL-
ESTINE

City of the Jebusites, The.
See: JERUSALEM, PAL-
ESTINE

City of the Kings, The. See:
CASHEL, IRELAND; LIMA,
PERU

City of the Kralitz Bible, The.
See: KRALICKA, CZECH-
OSLOVAKIA

City of the Lagoons, The.
See: VENICE, ITALY

City of the Leaning Tower,
The. See: PISA, ITALY

City of the Locrian Code, The.
See: LOCRI, ITALY

City of the Maid, The. See:
ORLEANS, FRANCE

City of the Massacre, The.
See: CAWNPORE, INDIA

City of the Mines, The. See:

IGLESIAS, SARDINIA

City of the Pool of Immortality, The. See: AMRITSAR,
INDIA

City of the Prophet, The. See:
MEDINA, SAUDI ARABIA

City of the Reef, The. See:
PERNAMBUCO, BRAZIL

City of the Refuge, The. See:
MEDINA, SAUDI ARABIA

City of the Simple, The. See:
GHEEL, BELGIUM

City of the Storks, The. See:
TIMBUKTU, AFRICA

City of the Sun, The. See:
BAALBEK, LEBANON; CUZ-
CO, PERU; HELIOPOLIS,
EGYPT

City of the Synod, The. See:
GANGRA, TURKEY

City of the Three Kings, The.
See: COLOGNE, GERMANY

City of the Tribes, The. See:
GALWAY, IRELAND

City of the Turones, The. See:
TOURS, FRANCE

City of the Violated Treaty,
The. See: LIMERICK,
IRELAND

City of the Violet Crown, The.
See: ATHENS, GREECE

City of the West, The. See:
GLASGOW, SCOTLAND

City of Venus, The. See:
MELOS, GREECE

City of Victory, The. See:
CAIRO, EGYPT

City of Wallenstein's Death,
The. See: CHEB, CZECH-
OSLOVAKIA

City of Zaleucus, The. See:
LOCRI, ITALY

City of Zeus, The. See:
THEBES, EGYPT

City on the Golden Horn, The.
See: ISTANBUL, TURKEY

City Under the Ice, The. See:
CAMP CENTURY, GREEN-
LAND

City Where Kings Are Crowned,
The. See: TRONDHEIM,
NORWAY

Ciudad Trujillo. See: SANTO
DOMINGO, HISPANIOLA

Civilization, The Cradle of.
See: MEDITERRANEAN
SEA

Civilization, The Cradle of
Chinese. See: WEI RIVER
VALLEY

CLAIRVAUX, FRANCE. Ham-
let of Aube dept. , N. W.
France 40 mi. S. E. of
Troyes

The Home of St. Bernard

CLARENDON PARK, ENGLAND.
Parish of S. Wiltshire,
England, 2 mi. S. E. of
Salisbury

The Home of the Constitu-
tions

CLEMENTE. See: SAN
CLEMENTE, CALIFORNIA

Clinton's Big Ditch. See:
ERIE CANAL

Clinton's Ditch. See:

ERIE CANAL

Cloth Hall, The Home of the.
See: IEPER (YPRES), BEL-
GIUM

Cloth of Gold. See: Field of
the Cloth of Gold, The

CLYDESDALE, SCOTLAND.
Valley of upper Clyde R. ,
S. Scotland, about 50 mi.
long

The Home of the Clydesdale
Horse

Coarse Pounded Corn, The
River of. See: CHICKA-
HOMINY RIVER

Coast. See: Barbary Coast,
The; Rocket Coast, The

Coast, Australia's Sunshine.
See: NOOSA HEADS

Coast, The Azure. See: RIV-
IERA, THE

Coast, The Grain. See: LI-
BERIA

Coast, The Kru. See: LI-
BERIA

Coast, The Tobacco. See:
CHESAPEAKE BAY

Coast of the East, The Barbary.
See: SANDS STREET, BROOK-
LYN, N. Y.

Coast of the Rising Sun, The.
See: RIVIERA, THE

Coast of the Setting Sun, The.
See: RIVIERA, THE

Coat Rack, The. See: SYD-
NEY HARBOR BRIDGE

Cocagne. See: LONDON,
ENGLAND

COCAGNE (COCAIGNE), THE
LAND OF.   Fabled land
of happiness "overflowing
with milk and honey"

Lubberland

Cockpit of Europe, The.   See:
BELGIUM

COD, CAPE.   Sandy peninsula
of S. E. Massachusetts lying
between Cape Cod Bay and
Atlantic Ocean

The Cape

Code, The City of the Locrian.
See: LOCRI, ITALY

Codfish, The Home of the.
See: GRAND BANK

Coffee, The City of. See:
SANTOS, BRAZIL

Coffee Plant, The Home of
the. See: KAFA (KAFFA)

COGNAC, FRANCE.   Town of
S. W. France sit. on Char-
ente R. 23 mi. N. N. W. of
Angou Lême

The Town of Brandy

COLCHESTER, ENGLAND.
City of Essex, England,
on Colne R. 52 mi. N. E.
of London

The Royal Town of Cunobelin

COLCHIS.   Ancient country on
Black Sea S. of Caucasus
Mountains

Medea's Home

Cold Estotiland.   See: ES-
TOTILAND

COLDBATH FIELDS.   Prison

of London, England

The English Bastille

COLLEGE.   See: TRINITY
COLLEGE

COLLEGIATE CHURCH OF ST.
PETER.   National sanctuary
and burying place sit. in the
borough of Westminster, London, England

The Pantheon of the British
Westminster Abbey

COLOGNE, GERMANY.   City
of North Rhine-Westphalia,
W. Germany, on Rhine R.
325 mi. S. W. of Berlin

The City of the Three Kings
The Rome of the North

COLONY.   See: PLYMOUTH
COLONY

Colony, The City of the All-
British. See: LLOYDMIN-
STER, CANADA

Colony, The Fourteenth.   See:
CANADA

Colony, The Old.   See: PLY-
MOUTH COLONY

COLUMBIA RIVER.   R. of
S. W. Canada and N. W.
United States flowing from
Columbia Lake, B. C. , to
Pacific Ocean; about 1270
mi. long

The Achilles of Rivers

Columbus, The Birthplace of.
See: GENOA, ITALY

COLUMBUS CIRCLE, NEW
YORK CITY.   Area of New
York City

The Hub of New York

Columbus's First Landfall.
See: SAN SALVADOR
(WATLINGS ISLAND)

Columbus's First Settlement.
See: ISABELA, NORTH
DOMINICAN REPUBLIC

Columns. See: Temple of
a Thousand Columns, The

Commercial Highway of Southeast China, The Great.
See: SI-KIANG

Communes. See: Seven
Communes, The; Thirteen
Communes, The

COMPOSTELA. See: SAN-
TIAGO DE COMPOSTELA

CONDOM, FRANCE. Com-
mune, Gers dept., S. W.
France, 20 mi. S. W. of
Agen

The Home of Armagnac
Brandy

Confederacy, The Backbone
of the. See: MISSISSIPPI
RIVER

Conference, The Place of the
Big Three. See: YALTA,
RUSSIA

CONGO, REPUBLIC OF THE.
African republic of the French
Community, Central Africa

The Paradise of the World

CONGO RIVER. Second largest R. in Africa and fourth
in size in the world; about
2900 mi. long

Pillar River

CONJEEVERAM, INDIA. Town
of Madras, India, 40 mi.

W. S. W. of city of Madras

The Benares of the South

Connemara, The English. See:
AGAR-TOWN, ENGLAND

Constantinople, Little. See:
KERCH, U. S. S. R.

Constitutions, The Home of the.
See: CLARENDON PARK,
ENGLAND

Continent, The Backbone of the
American. See: ROCKY
MOUNTAINS

Continent, The Dark. See:
AFRICA

Continent, The Extremity of
the. See: GOOD HOPE,
CAPE OF

Continent, Half a. See: BRA-
ZIL

Continent, The Island. See:
AUSTRALIA

Continent, The Roof of the.
See: ROCKY MOUNTAINS

Continent, The Sunken. See:
ATLANTIS

Continent of the Pacific, The
Sunken. See: LEMURIA

CONTINENTAL DIVIDE. El-
evated ridge in Rocky Mts.
of N. America separating
Pacific Ocean from Atlantic
Ocean tributaries

The Great Divide

Cooperative Movement, The
Birthplace of the. See:
ROCHDALE, ENGLAND

COPENHAGEN, DENMARK.

Industrial and commercial city and seaport on E. coast of Denmark

The Athens of the North

CORDOVA, SPAIN. City of Spain sit. on Guadalquivir R. 73 mi. E. N. E. of Seville

The Athens of the West

CORFU. Ionian island sit. in Ionian Sea off coast of S. W. Albania and N. W. of Greece

The Lovely Isle

CORINTH, GREECE. Ancient city of Greece located at southern extremity of Isthmus of Corinth 3 mi. S. W. of modern town of Corinth

The Light of Greece
The Paris of the Ancient World

CORK, IRELAND. City of S. W. Ireland, Munster province, at mouth of Lee R.

The Athens of Ireland
The Drisheen City

Corn, The River of Coarse Pounded. See: CHICKA-HOMINY RIVER

Corner. See: Dead Man's Corner

Corners. See Four Corners

CORNO, MONTE. Peak in Apennine Mts. , Teramo province, Italy, 9585 ft. high

The Great Rock of Italy

CORNWALL, ENGLAND. Maritime county sit. in extreme S. W. part of England

Land's End

Coronation City, The. See: VLADIMIR, U. S. S. R.

CORREGIDOR ISLAND. Rocky island at entrance to Manila Bay, Luzon, P. I.

The Gibraltar of the East
The Rock
The Rock of Corregidor

Cortes, The Sea of. See: CALIFORNIA, GULF OF

Côte d'Azur. Mediterranean coast of France, especially its E. end, so called because of deep blue color of the sea; part of the Riviera

Council of Trent, The Seat of the. See: TRENTO, ITALY

Counties. See: Home Counties, The; Six Counties, The

Countries. See: Low Countries, The

Country. See: Black Country, The; Fen Country, The

Country, The Eagle's. See: ALBANIA

Country, O'Donnell's. See: DONEGAL, IRELAND

Country, The Old. See: AUSTRALIA; ENGLAND; GREAT BRITAIN

Country, The Owld. See: IRELAND (EIRE)

Country Between Two Rivers, The. See: MESOPOTAMIA

Country in Search of a Future, The. See: MALAY
ARCHIPELAGO

Country of Paradoxes, The.
See: NETHERLANDS, THE
(HOLLAND)

Country of the Blacks, The.
See: SUDAN, AFRICA

Country of the Kanarese, The.
See: CARNATIC, INDIA

Country of the Olympic Games,
The. See: ELIS

Court, Venus's. See: HÖRSEL
BERGE, GERMANY

COVENTRY, ENGLAND. City
and borough of Warwick-
shire, central England, near
Avon R. 18 mi. E. S. E. of
Birmingham

The Home of Lady Godiva
The Town of Godiva's Ride

Cradle of American Liberty,
The. See: FANEUIL
HALL; INDEPENDENCE
HALL

Cradle of Chinese Civilization,
The. See: WEI RIVER
VALLEY

Cradle of Civilization, The.
See: MEDITERRANEAN
SEA

Cradle of Liberty, The. See:
FANEUIL HALL

Cradle of Swiss Freedom, The.
See: SCHWYZ, SWITZER-
LAND

Cradle of the Penitentiary, The.
Appellation of Walnut Street
Jail at Philadelphia, Pa.

Cradle of the Reformation, The.

See: WITTENBERG, GER-
MANY

Crag, The King's. See: FIFE,
SCOTLAND

CREMONA, ITALY. City of
Lombardy, N. Italy, sit.
on Po R. 49 mi. E. S. E. of
Milan

The Violin City

Crimson Carpet, Tangipahoa of
the. See: TANGIPAHOA

Crocodiles, The River of. See:
LIMPOPU RIVER

Croesus of English Abbeys,
The. See: RAMSAY ABBEY

CROMER BAY. Inlet on coast
of North Sea near Norfolk,
England

Devil's Throat

Crossroads, The Nation at the.
See: BURMA

Crossroads of South America,
The Heroin. See: PARA-
GUAY

Crossroads of the Far East,
The. See: SINGAPORE,
ASIA

Crown, The City of the Violet.
See: ATHENS, GREECE

Crown of Ionia, The. See:
SMYRNA, ASIA MINOR

Crown of the East, The. See:
ANTIOCH, ASIA MINOR

Crowned, The City Where Kings
Are. See: TRONDHEIM,
NORWAY

Crusoe's Island. See: MAS A
TIERRA

Crystal Hills, The. See:
WHITE MOUNTAINS

CUBA. Largest island of
West Indies sit. in Atlantic Ocean S. of Florida
peninsula and E. of Yuca-
tán peninsula of Mexico

The Ever-Faithful Isle
The Flower of Islands
The Hero of Africa
The Key of the Gulf
The Pearl of the Antibes
The Pearl of the Antilles
The Queen-Faithful Isle

Cudgel, The Town of the.
See: SHILLELAGH, IRE-
LAND

CUMAE, ITALY. Ancient
town of Campania, Italy,
on coast W. of Naples;
oldest Greek colony in
Italy or Sicily

The Home of the Sibyl

Cunobelin, The Royal Town
of. See: COLCHESTER,
ENGLAND

CURES, ITALY. Ancient
town of Latium, Italy, N. E.
of Rome

The Home of the Sabines

CUTHAH, BABYLONIA. An-
cient city of Babylonia de-
voted to worship of Nergal,
ruler of Aralu, the abode
of the dead

The City of Nergal

CUZCO, PERU. City of
Peru sit. about 350 mi.
S. E. of Lima

The Archaeological Cap-
ital of South America

The City of the Sun
The Holy City

Czar's Village, The. See:
Tsar's Village, The

-D-

Dabney, The Isle De. See:
MADEIRA

DACCA, BANGLADESH. Cap-
ital city of Bangladesh and
district of Dacca, E. Bengal

The Eastern Capital

Dagon, The City of. See: GA-
ZA, PALESTINE

DAMANHUR, UNITED ARAB
REPUBLIC. Capital of
Beheira province, United
Arab Republic, sit. 38 mi.
E. S. E. of Alexandria

The Town of Horus

DAMASCUS, SYRIA. Capital
and chief city of Syria, sit.
on both banks of Barada R.,
55 mi. S. E. of Beirut (Bey-
routh), Lebanon

The City of Mosques
The Holy City

DANUBE RIVER. Second largest traffic artery of Europe,
flowing from Black Forest
region, Germany, to Black
Sea; approx. 1725 mi. long

The Beautiful Blue Danube
The Blue Danube

Dardanelles of the New World,
The. See: DETROIT RIV-
ER

Dards, The Home of the. See:

HUNZA, INDIA

Dark Continent, The. See:
AFRICA

Darkest Africa. See: AF-
RICA

Darkness, The Sea Of. See:
ATLANTIC OCEAN

DARTFORD, ENGLAND. Ur-
ban district of Kent, S. E.
England, sit. on Darent R.
15 mi. E. S. E. of London

Wat Tyler's Bailiwick

DASHT-I-KAVIR. Desert
sit. in N. central Iran

The Great Salt Desert

Daughter of Rome, The Beautiful. See: FLORENCE,
ITALY

DAVID. See: CAMP DAVID,
MARYLAND

David, The City of. See:
JERUSALEM, PALESTINE;
TEMPLE HILL; ZION, PAL-
ESTINE

David, The Home of. See:
BETHLEHEM, PALESTINE

DAVIDSON, MOUNT. Peak in
Storey County, W. Nevada;
Site of Virginia City and
Comstock Lode

The Mountain of Silver

De Dabney, The Isle. See:
MADEIRA

DE JANEIRO. See: RIO DE
JANEIRO, BRAZIL

DE SANTA MARIA. See:
PUERTO DE SANTA

MARIA, SPAIN

Dead, The City of the. See:
VERACRUZ, MEXICO

Dead Man's Corner. Part of
Antietam, Maryland, battlefield in American Civil War

DEAD SEA. Salt lake sit. on
boundary of Palestine and
Transjordan

The Lake of Lot
The Salt Sea
The Sea of Lot

Death, The City of the Black.
See: TULLE, FRANCE

Death, The City of Wallenstein's. See: CHEB,
CZECHOSLOVAKIA

Death's Door. Entrance from
Lake Michigan into Green
Bay, Wisconsin

Debatable Land, The. Region
on borders of England and
Scotland between the Esk
and Sark Rivers formerly
claimed by both kingdoms

Deccan, The Palmyra of the.
See: BIJAPUR, INDIA

DEEP. See: WHARTON DEEP

Deity, The House of. See:
BETHLEHEM, PALESTINE

DELPHI, GREECE. Town in
ancient Greece; site now occupied by Delphoi in the department of Phthiotis and
Phocis

The Home of the Oracle

Delta City, The. See: ALEX-
ANDRIA, EGYPT

Demeter, The City of. See:

ELEUSIS, GREECE

Denmark. See: Windsor of
Denmark, The

Denmark, The Orchard of.
See: FALSTER ISLAND

Denmark, The Wapping of.
See: HELSINGÖR (EL-
SINORE), DENMARK

Deportations, the City of.
See: LILLE, FRANCE

DERBENT, RUSSIA IN EU-
ROPE. Town of S. E. Dag-
estan, S. E. Soviet Russia
in Europe, sit. on Caspian
Sea 70 mi. S. E. of Mak-
hachkala

The Caspian Gates
The Derbent Gateway
The Iron Gates

Derby, The Home of the.
See: CHURCHILL DOWNS,
LOUISVILLE, KY; EPSOM
DOWNS, ENGLAND

Descender, The. See: JOR-
DAN RIVER

DESERT. See: ARABIAN
DESERT; GOBI DESERT

Desert. See: Great American
Desert, The

Desert, The Black. See:
KARA KUM

Desert, The Desert Within a.
See: TANEZROUFT

Desert, The Eastern. See:
ARABIAN DESERT

Desert, The Gem of the. See:
GRAAF REINET, SOUTH
AFRICA

Desert, The Great Salt. See:

DASHT-I-KAVIR

Desert Within a Desert, The.
See: TANEZROUFT

DESSAU, GERMANY. City of
Anhalt, Germany, sit. on
Mulde R. 71 mi. S. W. of
Berlin

The City of Albert the Bear

Destiny, The Land of. See:
IRELAND (EIRE)

Detroit, The Wall Street of.
See: GRISWOLD STREET,
DETROIT, MICH.

DETROIT RIVER. River connecting Lake St. Clair with
Lake Erie, central N. America; 31 mi. long

The Dardanelles of the New
World

Devil's Backbone, The. See:
NATCHEZ TRACE

Devil's Island, America's. See:
ALCATRAZ ISLAND

Devil's Islands (Isles du Diable). See: SAFETY IS-
LANDS (ISLES DU SALUT)

Devil's Throat. See: CROM-
ER BAY

Devil's Wall, The. Southern
portion of Roman fortifica-
tions called the Pfahlgraben,
in England; built c. 70 A.D.

DHAHRAN, SAUDI ARABIA.
City of al-Hasa district, E.
Saudi Arabia, near W. coast
of Persian Gulf

The Oil Capital of the World

Diable, Isles du (Devil's Islands). See: SAFETY ISLANDS (ISLES DU SALUT)

DIAMOND, CAPE. Promontory at E. end of city of Quebec, Canada

The Gibraltar of the New World

Diamond Capital of the World, The. See: ANTWERP, BELGIUM

Diamond Center, The World's. See: KIMBERLEY, SOUTH AFRICA

Diamonds, The City of. See: GOLCONDA, INDIA

Diamonds, The Town of. See: KIMBERLEY, SOUTH AFRICA

Dido's City. See: CARTHAGE, NORTH AFRICA

Diets, The Mother of. See: WORMS, GERMANY

Dike. See: Grahame's Dike

Dismal Swamp, The Lake of the. See: DRUMMOND, LAKE

Distant South, The. See: VIET NAM

District. See: Lake District, The; Shoestring District, The

Ditch. See: Fleet Ditch, The

Ditch, The Big. See: PANAMA CANAL

Ditch, The Busy. See: PANAMA CANAL

Ditch, Clinton's. See: ERIE CANAL

Ditch, Clinton's Big. See: ERIE CANAL

Ditch, The World. See: SUEZ CANAL

DIVIDE. See: CONTINENTAL DIVIDE

Divide, The Great. See: CONTINENTAL DIVIDE

Divided Dragon, The. See: VIET NAM

Dividing Line, The. See: YALU RIVER

Dixie. Land and states S. of Mason-Dixon Line, N. America

Doctors of the Notariate, The Fountain of. See: FLORENCE, ITALY

DODECANESE ISLANDS. Group of about 50 islands and islets sit. in S. E. part of Aegean Sea off coast of Asiatic Turkey

Twelve Islands

Dogs, The Islands of. See: CANARY ISLANDS

Doldrums. Ocean regions near the equator, characterized by calms or light winds

Dolorosa. See: Via Dolorosa (The Way of Pain)

DOME. See: TEAPOT DOME, WYOMING

Dome of the Rock, The. Mosque of Omar, sit. in Jerusalem, Palestine

DOMINGO. See: SANTO DO-

MINGO, HISPANIOLA

DOMINICAN REPUBLIC. Republic in E. two thirds of Hispaniola Island, West Indies

The Black Republic

DONEGAL, IRELAND. Northwest county of Republic of Ireland bounded on N. W. by Atlantic Ocean

O'Donnell's Country

DONNYBROOK, IRELAND. Suburb of city of Dublin, E. Ireland; scene of annual fair

The Home of Fighting for Fun

Door. See: Death's Door

Dorado. See: El Dorado

DORCHESTER, ENGLAND. County town of Dorsetshire, England, sit. on Frome R. 6 mi. N. of English Channel

The Home of the Bloody Assizes

DOUAI, FRANCE. City of Nord dept., N. France, sit. on Scarpe R., 19 mi. S. of Lille

The Birthplace of the Douai Bible

Doubt, The River of. See: RIO ROOSEVELT

DOUCATO, CAPE. Promontory on island of Leukas, Ionian Sea

Sappho's Leap

DOVER, ENGLAND. Seaport and municipal borough of Kent County, England, 76 mi. E. S. E. of London

The Chief of the Cinque Ports

Down Under. See: AUSTRALIA

Downing Street. See: No. 10 Downing Street

DOWNS. See: CHURCHILL DOWNS, LOUISVILLE, KY.; EPSOM DOWNS, ENGLAND

DRACHENFELS. Rock, one of the Siebengebirge, sit. on E. bank of Rhine R., S. of Bonn, W. Germany; 1053 ft. high

The Dragon's Rock

Dragon, The Divided. See: VIET NAM

Dragon, The Hungry. See: CHINA

Dragon Lizard, The Home of the. See: KOMODO

Dragons, The Land of Many. See: VIET NAM

Dragon's Rock, The. See: DRACHENFELS

Dreams, The City of a Million. See: VIENNA, AUSTRIA

DRESDEN, GERMANY. City of Saxony, Germany, on Elbe R. 63 mi. E. S. E. of Leipzig

The Elbe Florence
The Florence of the North
The German Florence
The Home of Dresden China

Drisheen City, The. See:

CORK, IRELAND

Druids, The Place of the. See:
ANGLESEY, ENGLAND

DRUMMOND, LAKE. Lake in
midst of Dismal Swamp, S. E.
Virginia near N. Carolina
border

The Lake of the Dismal
Swamp

DU SALUT, ISLES. See: SAFE-
TY ISLANDS (ISLES DU SA-
LUT)

Dual Monarchy, The. See:
AUSTRIA-HUNGARY

DUBLIN, IRELAND. Capital,
county borough and seaport
of Republic of Ireland sit.
on Dublin Bay, inlet of Irish
Sea

The Town of the Ford of the
Hurdles

DUMBARTON, ROCK OF. Twin-
peaked hill sit. on bank of
Clyde R. , Scotland, near its
junction with Leven R.

The Gibraltar of Scotland

Dumping Ground of Europe, The.
See: GREAT BRITAIN

Dunes, Church on the. See:
DUNKIRK, FRANCE

DUNKIRK, FRANCE. Seaport
in department of Nord, France,
on Strait of Dover 28 mi.
N. E. of Calais

Church on the Dunes

Dust Bowl, The. Area of
Great Plains lying between
Mississippi R. and Rocky
Mts. , Canada to Mexico

Dust Stream, The. See:
STAUBBACH

Dusting City, The Knuckle.
See: LIVERPOOL, ENG-
LAND

Dutch Paradise, The. See:
GELDERLAND, E. NETHER-
LANDS

Dutch Seas, The Key of the.
See: FLUSHING, THE
NETHERLANDS

- E -

Eagles. See: House With the
Eagles, The

Eagle's Country, The. See:
ALBANIA

Earl's Seat, The. Highest
point of Lennox Hills, sit.
in Dumbarton and Stir-
ling Counties, S. W. cen-
tral Scotland, 1894 ft.
high

Early Chinese Agriculture, The
Home of. See: SHANSI,
CHINA

Earth, Hell on. See: ANDER-
SONVILLE PRISON

East. See: Niagara of the
East, The

East, The. Appellation of the
civilized Asiatic countries,
either ancient or modern

East, The. See also: Far
East, The; Near East, The

East, The Barbary Coast of
the. See: SANDS STREET,
BROOKLYN, N. Y.

East, The Britain of the Far.
See: JAPAN

East, The Crossroads of the
Far. See: SINGAPORE,
ASIA

East, The Crown of the. See:
ANTIOCH, ASIA MINOR

East, The Gibraltar of the.
See: ADEN, S. W. ARABIA;
CORREGIDOR ISLAND

East, the Glory of the. See:
PERSEPOLIS, PERSIA

East, Miami Beach. See:
TEL AVIV, ISRAEL

East, The Queen of the. See:
ANTIOCH, ASIA MINOR;
BATAVIA, NETHERLAND
INDIES

East, The Sick Man of the.
See: TURKEY

East, The Venice of the. See:
BANGKOK, THAILAND;
WUHSIEN, CHINA

East End. Part of London,
England, lying E. of Bank of
England; includes industrial
and shipping districts and
most of the poorer districts

East Side. East part of borough of Manhattan, New York
City, particularly below
14th St. (Lower East Side);
residence mostly of poorer
classes

East Valley of Norway, The.
See: OSTERDAL

Eastern America, The Rooftop
of. See: GREAT SMOKY
MOUNTAINS

Eastern Archipelago, The

Queen of the. See: JAVA

Eastern Capital, The. See:
DACCA, BANGLADESH;
TOKYO, JAPAN

Eastern Desert, The. See:
ARABIAN DESERT

Eastern Europe, The Paris of.
See: VIENNA, AUSTRIA

EDAM, THE NETHERLANDS.
Town of North Holland province, the Netherlands, 15
mi. N. E. of Amsterdam

The City of Cheese

Eden of America, The. See:
AQUIDNECK ISLAND

Eden of Germany, The. See:
BADEN, AUSTRIA

EDINBURGH, SCOTLAND.
City and royal burgh, cap-
ital of Scotland sit. 40 mi.
E. of Glasgow

The Athens of the North
Auld Reekie
The Castle of Maidens
The City of Palaces
The Maiden Town
The Modern Athens
The Mountain of Sorrow
The Northern Athens
The Queen of the North

EGYPT. Country of Africa
and Asia coextensive with
the United Arab Republic

The Granary of the Roman
World
The Land of Bondage
The Land of the Pharaohs
Sphinxland

Egyptian Resort City, The.
See: ASWAN, EGYPT

1883, The Moonshot of. See:

BROOKLYN BRIDGE

Eighth Wonder of the World,
The. See: BROOKLYN
BRIDGE

EILANDEN. See: OVERIGE
EILANDEN

EINBECK, WEST GERMANY.
Town of Lower Saxony State,
W. Germany, sit. on Ulm
R. , 40 mi. S. of Hannover

The Home of Bock Beer

EINSIEDELN, SWITZERLAND.
Town in canton of Schwyz,
Switzerland, sit. on Alp-
bach R. 26 mi. S. E. of
Zurich

The Loretto of Switzerland
The Place of the Hermits

EIRE. See: IRELAND (EIRE)

El Dorado. Legendary fabulously wealthy city or area
supposed to exist in N. part
of South America

EL KHAZNA. Temple located
in ancient city of Petra, N.
Arabia

The Treasury of Pharaoh

EL MANDEB. See: BAB EL
MANDEB

El Puerto (The Port). See:
PUERTO DE SANTA MARIA,
SPAIN

EL QASR, EGYPT. Town of
Dakhla Oasis, S. Desert
province, Egypt

The Castle Stronghold

El Salvador, The Lighthouse of.
See: IZALCO

Elbe Florence, The. See:
DRESDEN, GERMANY

ELBERFELD, GERMANY. For-
mer city of Düsseldorf govt.
district, Rhine province,
Prussia, Germany; now part
of Wuppertal

The Manchester of Prussia

Elephant, The Land of the
White. See: THAILAND

Elephant Tusk, The. See: IN-
DIANOLA PEAK

Eleusinian Mysteries, The Seat
of the. See: ELEUSIS,
GREECE

ELEUSIS, GREECE. Village
with ruins of ancient city,
about 14 mi. N. W. of Athens, Greece

The City of Demeter
The Seat of the Eleusinian
Mysteries

Eli and Samuel, The Home of.
See: SHILOH, PALESTINE

ELIS. Ancient kingdom in N. W.
Peloponnesus, Greece, on
Ionian Sea

The Country of the Olympic
Games
The Holy Land

ELLIS ISLAND. Island sit. in
upper bay of New York harbor 1 mi. S. W. of Battery
Park

Gibbet Island
Oyster Island

ELSINORE. See: HELSINGÖR
(ELSINORE), DENMARK

ELSTER. See: SCHWARZE
ELSTER

Elster, The Black. See:
SCHWARZE ELSTER

Emerald Island of the West,
The. See: MONTSERRAT

Emerald Isle, The. See: IRE-
LAND (EIRE)

Emerald of Europe, The. See:
IRELAND (EIRE)

EMPIRE. See: HOLY ROMAN
EMPIRE

Empire, The Celestial. See:
CHINA

Empire, The Heart of the.
See: MOSCOW, RUSSIA

Empire, The Holy. See:
HOLY ROMAN EMPIRE

Empire, The Jewel of the
British. See: INDIA

Empire, The Roman. See:
HOLY ROMAN EMPIRE

Empire of Bees, The. See:
HYBLA, SICILY

Empire of the Islands, The.
See: MALAY ARCHIPEL-
AGO

Empire of the Seven Rivers,
The. See: PUNJAB, IN-
DIA

Empire of the West, The.
See: HOLY ROMAN EM-
PIRE

Emporium of South American
Trade, The Sixteenth-Cen-
tury. See: PORTO BELLO,
PANAMA

Empress Josephine's Island.
See: MARTINIQUE

EN-JOSAS. See: JOUY-EN-
JOSAS, FRANCE

Enchanted City, The. See:
PETRA, ARABIA

End. See: East End

END. See: WEST END

End, Land's. See: CORN-
WALL, ENGLAND; FIN-
ISTERRE, CAPE

Enderby Quadrant, The. For-
merly quarter section of
Antarctic continent between
Greenwich Meridian and 90°
E.; now chiefly Queen Maud
Land, Enderby Land, Mac-
Robertson Coast and Leo-
pold and Astrid Coast

ENGLAND. S. part of island
of Great Britain, excluding
Wales; largest division of
United Kingdom of Great
Britain and Ireland

Albion
The Land of the Rose
Merrie England
Merry England
The Mother of Parliaments
The Old Country
The Ringing Island
The Sea-Girt Isle
The Tight Little Island
The Tight Little Isle

ENGLAND. See also: GREAT
BRITAIN

England, The Achilles' Heel of.
See: IRELAND (EIRE)

England, The Garden of. See:
HEREFORD, ENGLAND;
KENT, ENGLAND; WIGHT,
ISLE OF; WORCESTERSHIRE,
ENGLAND

England, The Heart of. See:
WARWICKSHIRE

England, Little. See: BAR-
BADOS

England, The Top of New. See:
WHITE MOUNTAINS

English Abbeys, The Croesus of.
See: RAMSAY ABBEY

English Bastille, The. See:
COLDBATH FIELDS

ENGLISH CHANNEL. Strait
between S. England and N.
France

The Channel
The Silver Streak
The Sleeve (La Manche)

English Connemara, The. See:
AGAR-TOWN, ENGLAND

English Justice, The Birthplace
of. See: RUNNYMEDE,
ENGLAND

English Pale, The. Those
parts of Ireland in which
English law was acknowl-
edged: Dublin, Meath, Carlow, Kilkenny and Louth

English Switzerland, The.
Neighborhood of Ilfracombe,
Lynton and Lynmouth, N.
Devon, England

English Town in South Africa,
The Most. See: GRAHAMS-
TOWN, SOUTH AFRICA

English Watering Places, The
Queen of. See: SCARBOR-
OUGH, ENGLAND

Entrance, Hell's. See: AVER-
NUS, LAKE

EPSOM, ENGLAND. Urban district of Surrey, S. England,
on edge of Banstead Downs,
15 mi. S. S. W. of London

The Home of Epsom Salts

EPSOM DOWNS, ENGLAND.
Racetrack near Epsom, S.
England, 15 mi. S. S. W. of
London

The Home of the Derby

Epsom Salts, The Home of.
See: EPSOM, ENGLAND

EQUATORIAL AFRICA. Por-
tion of central Africa, second largest continent on the
globe

The New World of Tomorrow

EQUATORIAL AFRICA. See
also: AFRICA; SOUTH AF-
RICA

ERIE, LAKE. See: Great
Lakes, The

ERIE CANAL. Canal, 363 mi.
long, extending from Buffalo,
N. Y. , on Lake Erie to Al-
bany, N. Y. , on Hudson R.

Clinton's Big Ditch
Clinton's Ditch

Erin. See: IRELAND (EIRE)

Erin, The Garden of. See:
CARLOW, IRELAND

ERYTHRAE, LYDIA. Ancient
city of Lydia on coast of
peninsula opposite island of
Chios

The Home of Herophile

Escape-Proof Prison, America's First. See: ALCA-
TRAZ ISLAND

Española, La (Little Spain).
See: HISPANIOLA

ESSLINGEN, WEST GERMANY.

City of Baden-Württemberg
state, W. Germany, 6 mi.
E. S. E. of Stuttgart

The Home of Neckar Wine

ESTOTILAND. Mythical land
in America located by old
geographers in Newfound-
land, Labrador, and Canada
E. of Hudson Bay

Cold Estotiland

Eternal City, The. See:
ROME, ITALY

ETHIOPIA (ABYSSINIA). Constitutional monarchy of N. E.
Africa W. of Red Sea

The Land of Prester John
The Montenegro of Africa

Ethiopia, The Olympus of.
See: KILIMA, MOUNT

ETIENNE. See: SAINT-
ETIENNE, FRANCE

ETNA, MOUNT. Volcano in
N. E. Sicily near the coast;
10,741 ft. high

The Mountain of Fire

Eurasian Plain, The. Great
central plateau of Europe
and Asia

Europe. See: Powder Keg of
Europe, The

Europe, The China of. See:
AUSTRIA-HUNGARY

Europe, The Cockpit of. See:
BELGIUM

Europe, The Dumping Ground
of. See: GREAT BRITAIN

Europe, The Emerald of.

See: IRELAND (EIRE)

Europe, The Garden of. See:
BELGIUM; ITALY

Europe, The Gold Mine of.
See: TRANSYLVANIA

Europe, The Granary of. See:
SICILY

Europe, The Intellectual Center of Protestant. See:
GENEVA, SWITZERLAND

Europe, The Oldest State in.
See: SAN MARINO

Europe, The Paris of Eastern.
See: VIENNA, AUSTRIA

Europe, The Playground of.
See: SWITZERLAND

Europe, The Sick Man of.
See: TURKEY

Europe, The Westernmost
Point of. See: SAINT VIN-
CENT, CAPE

Europe, A Window into. See:
LENINGRAD, U. S. S. R.

European Saratoga, The. See:
BADEN-BADEN, GERMANY

Europe's Africa. See: BALKAN
MOUNTAINS

Evangeline, The Home of. See:
GRAND PRE, NOVA SCOTIA

Ever-Faithful Isle, The. See:
CUBA

Ever-Loyal City, The. See:
OXFORD, ENGLAND

EVEREST, MOUNT. Highest
known mountain in the world
(29,002 ft.) sit. in Himalaya Mts. on frontier

between Nepal and Tibet

The Goddess-Mother
The Goddess Mother of the
  World
The Goddess Mountain
The Jealous Goddess
The Mysterious Mountain
Peak XV
The Summit of the World
The Top of the World

Everlasting Fire, The House
  of. See: HALEMAUMAU

Exactress of Gold, The. See:
  BABYLON

Extremes, The Street of. See:
  WALL STREET, NEW YORK
  CITY

Extremity of the Continent,
  The. See: GOOD HOPE,
  CAPE OF

Eye of Greece, The. See:
  ATHENS, GREECE

Eye of the Baltic, The. See:
  GOTLAND

- F -

Face, The Great Stone. See:
  CANNON MOUNTAIN

Fair City, The. See: PERTH,
  SCOTLAND

Faithful Isle, The Ever. See:
  CUBA

Faithful Isle, The Queen. See:
  CUBA

Fall, Fall's. See: TEAPOT
  DOME, WYOMING

Fall, The Golden. See:
  GULLFOSS

Falls, The Seven. See GUAIRA

Fall's Fall. See: TEAPOT
  DOME, WYOMING

Falls of Mexico, The Niagara.
  See: JUANACATLAN

FALSTER ISLAND. Island
  forming part of Denmark,
  lying in Baltic Sea S. W. of
  Möen and E. of Lolland

  The Orchard of Denmark

FANEUIL HALL. Public hall
  and market in Boston, Mass.
  sit. in Dock Square; meeting place for American pa-
  triots during Revolutionary
  period

  The Cradle of American
    Liberty
  The Cradle of Liberty

Far East, The. The Orient.
  Appellations of the countries
  of E. Asia bordering on the
  Pacific Ocean

Far East, The. See also:
  East, The; Near East, The

Far East, The Britain of the.
  See: JAPAN

Far East, The Crossroads of
  the. See: SINGAPORE,
  ASIA

Far West, The. See: MOR-
  OCCO

Farthest North, The. See:
  HAMMERFEST, NORWAY

Farthest South, The. See:
  HOWE, CAPE

Farthest South, America's.
  See: SABLE, CAPE

Farthest South, Australia's.
  See: WILSON'S PRO-
  MONTORY

Farthest West, Africa's. See: VERDE, CAPE

Farthest West, Australia's. See: STEEP POINT

Farthest West, Norway's. See: STEINSOY, NORWAY

Fascist Protection Wall, The Anti-. See: BERLIN WALL

Fashion and Pleasure, The London of. See: WEST END

Fat, The. See: BOLOGNA, ITALY

Fat Woman, The. See: IXTA-CIHUATL

Father of Waters, The. See: IRRAWADDY RIVER; MISSISSIPPI RIVER

Father Tiber. See: TIBER RIVER

Fatherland, The. See: GERMANY

FATSHAN, CHINA. City of central Kwangtung province, S. E. China

The Birmingham of China

Feet of Gold, The Camel with the. See: LIBYA (LIBIA)

Fen Country, The. The Fens. Bedford Level. Part of E. England formerly characterized by many swamps

Fens, The. Swampy region of Cambridge, Mass.

Fens, The. See also: Fen Country, The

FERNEY-VOLTAIRE, FRANCE.

Town of Ain dept. , E. France, on shore of Lake Geneva, 4 mi. from city of Geneva

The Town of Watchmakers

Fetish, The Land of the. See: AFRICA

FIELD. See: RANDOLPH FIELD, TEXAS

Field of Blood, The. Battlefield of Cannae, S. E. Italy, where Hannibal defeated Roman army in 216 B. C.

Field of Mourning, The. Battle field near Aragon, N. E. Spain; scene of military encounter between Christians and Moors July 17, 1134

Field of the Cloth of Gold, The. Plain near Ardes, Pas-de-Calis, France; meeting place of Francis I of France and Henry VIII of England, 1520

Field of the Forty Footsteps, The. Southampton Fields. Former meadow in London, England, where British Museum now stands

FIELDS. See: COLDBATH FIELDS

Fields, Southampton. See: Field of the Forty Footsteps, The

FIFE, SCOTLAND. County of E. Scotland sit. between firths of Tay and Forth

The King's Crag

Fighting for Fun, The Home of. See: DONNYBROOK, IRELAND

Fiji, Middle. See: LOMAI VITI

Finish Line, The. See: JOHN O' GROAT'S HOUSE

FINISTERRE, CAPE. High promontory at N.W. extremity of Spain on coast of La Coruña province

Land's End

FINLAND. Republic of N. Europe lying between lat. 60° and 70° N. and long. 20° and 32° E.

The Land of a Thousand Lakes

Fire, The House of Everlasting. See: HALEMAUMAU

Fire, The Mountain of. See: ETNA, MOUNT

First Escape-Proof Prison, America's. See: ALCATRAZ ISLAND

First Landfall, Columbus's. See: SAN SALVADOR (WATLINGS ISLAND)

First Settlement, Columbus's. See: ISABELA, NORTH DOMINICAN REPUBLIC

Fish Sea, The. See: AZOV, SEA OF

Fisherman's Paradise of the N. Atlantic, The. See: BLOCK ISLAND

Five Boroughs, The. Collective appellation of English boroughs of Derby, Leicester, Lincoln, Nottingham and Stamford

Five Rivers. See: PUNJAB, INDIA

Five Seas, The Port of. See: MOSCOW, RUSSIA

FLANNAN ISLANDS. Group of seven small islands in Atlantic Ocean in Outer Hebrides, W. of Scotland

The Seven Hunters

Fleet Ditch, The. Tidal stream which once flowed by W. wall of London, England

Floating Gardens, The Town of. See: XOCHIMILCO, MEXICO

Flora, The Metropolis of. See: ARANJUEZ, SPAIN

FLORENCE, ITALY. City of Firenze province, Tuscany, central Italy, sit. on Arno R. 146 mi. N.N.W. of Rome

The Beautiful Daughter of Rome
The City of Lilies
The Fountain of Doctors of the Notariate
The Gentle

Florence, The Elbe. See: DRESDEN, GERMANY

Florence, The German. See: DRESDEN, GERMANY

Florence of the North, The. See: DRESDEN, GERMANY

Flourishing Land of the Free, The. See: THAILAND

Flow of Wilderness, The. See: GALAPAGOS ISLANDS

Flower of Islands, The. See: CUBA

Flower of the Levant, The. See: ZANTE

Flowery Kingdom, The. See: CHINA

Flowery Kingdom, The Middle. See: CHINA

Flowery Land, The. See: CHINA

FLUSHING, THE NETHER-LANDS. Seaport of Zee-land province, the Nether-lands, sit. on Schelde es-tuary

The Key of the Dutch Seas

FLUSHING, THE NETHER-LANDS. See also: Cau-tionary Towns, The

Focal Point of the Reforma-tion, The. See: WITTEN-BERG, GERMANY

Footsteps. See: Field of the Forty Footsteps, The

Forbidden City, The. Walled enclosure in Peking, China, containing imperial palace; formerly closed to the pub-lic

Forbidden City, The. See al-so: LHASA (LASSA), TIBET; PEKING, CHINA

Forbidden Land, The. See: TIBET

Ford of the Hurdles, The Town of the. See: DUBLIN, IRE-LAND

Ford of the Ox, The. See: BOSPORUS

FOREST. See: SHERWOOD FOREST

Forest, Robin Hood's. See: SHERWOOD FOREST

Forest Cantons, The Four. Cantons of Lucerne, Schwyz, Uri and Unterwalden, Switz-erland, surrounding Lake of Lucerne

Forest Cantons, The Lake of the Four. See: LUCERNE, LAKE OF

Former Western White House, The. See: SAN CLEMENTE, CALIFORNIA

FORMOSA. See: TAIWAN (FORMOSA)

Forties. See: Roaring Forties, The

Fortunate Isles, The. See: CANARY ISLANDS

Forty Footsteps. See: Field of the Forty Footsteps, The

Fountain of Doctors of the No-tariate, The. See: FLOR-ENCE, ITALY

Fountain of Youth, The Island of the. See: BIMINI

Four. See: Forest Cantons, The Four

Four Cantons, The Lake of the. See: LUCERNE, LAKE OF

Four Corners. Locality in S.W. United States where boundaries of Colorado, New Mexico, Arizona and Utah come together

Four Forest Cantons, The Lake of the. See: LUCERNE, LAKE OF

Fourteenth Colony, The. See: CANADA

Fourth Part of a Province, The.

See: TETRARCHY OF LY-
SANIAS

Fowl, The Home of the Houdan. See: HOUDAN,
FRANCE

Fowl, The Town of the Orpington. See: ORPINGTON,
ENGLAND

FRANCE. Republic of W. central Europe sit. on continental mainland; separated
from England by English
Channel, Strait of Dover and
N. Sea

Marianne

France, The Garden of. See:
INDRE-ET-LOIRE; TOUR-
AINE

France, The Iron Gate of. See:
LONGWY, FRANCE

France, The Manchester of.
See: ROUEN, FRANCE

France, The Pittsburgh of.
See: SAINT-ETIENNE,
FRANCE

France, The Venice of. See:
AMIENS, FRANCE

FRASER ISLAND. Island off
S.E. coast of Queensland,
Australia

The Great Sandy Island

Free, The Flourishing Land of
the. See: THAILAND

Free and the Home of the
Brave, The Land of the.
See: UNITED STATES
OF AMERICA

Free Men, The Republic of.
See: TAIWAN (FORMOSA)

Freedom, The Cradle of Swiss.
See: SCHWYZ, SWITZER-
LAND

Freedom, The Sanctuary of.
See: OLD SOUTH CHURCH

French Quarter, The.   Area
of New Orleans, La.; once
red-light district

French Shore, The.   Neutra-
lized territory on S. and E.
coasts of Newfoundland extending from Cape May to
Cape St. John

French Switzerland, The.
Swiss cantons where French
is spoken (Geneva, Vaud,
Valais, Neuchâtel and parts
of Berne and Fribourg)

Friday Mosque, The. See:
MASJID JAMI

Friendly Islands, The. See:
TONGA ISLANDS

FRIJOLES, CANAL ZONE.
Town on E. shore of Gatun
Lake near course of Pana-
ma Canal, Canal Zone

Beantown

Frontier. See: Banal Frontier,
The

Frozen Time, The Land of.
See: ANTARCTICA

FUJI (FUJIYAMA), MOUNT.
Quiescent volcano in S.
Japan 70 mi. W.S.W. of
Tokyo

Mount Fuji
The Parnassus of Japan
The Sacred Mountain

Fun, The Home of Fighting for.
See: DONNYBROOK, IRELAND

FUNDY, BAY OF. Inlet of
Atlantic Ocean in S. E. Canada
extending from S. New Brunswick to Nova Scotia; known for
its swift tidal currents

The Bay of Tides

Fur Seal Islands, The. See:
PRIBLOFF ISLANDS

Future, The Country in Search of a.
See: MALAY ARCHIPELAGO

-G-

Gadsden Purchase, The. Tract
of land (29,670 sq. mi.) pur-
chased by U.S. from Mexico
for $10,000,000 in 1853 fol-
lowing negotiations conducted
by James Gadsden, American
minister to Mexico; now in
New Mexico and Arizona

GALAPAGOS ISLANDS. Island
group in Pacific Ocean on
equator, comprising Colón
province, Ecuador

The Flow of Wilderness
The Home of the Tortoise

GALILEE, SEA OF. Fresh
water lake of N. Israel,
Palestine, through which
Jordan R. flows

The Lake of Gennesaret
The Sea of Chinnereth
The Sea of Tiberias

Galloping Gertie. Bridge
between Tacoma and
Olympic Peninsula, Wash-
ington; collapsed during wind
storm of November 7, 1940

Gallows Hill. Hill near Salem,
Mass., where approximately 20
persons were hanged during
witchcraft persecutions of 1692

GALWAY, IRELAND. Municipal
borough and seaport at head of
Galway Bay, Ireland

The City of the Tribes

Games, The Country of the
Olympic. See: ELIS

Games, The Home of the Tailtean. See: TELLTOWN,
IRELAND

GANGES RIVER. Major riv-
er of Indian subcontinent
flowing from Uttar Pradesh
state, Republic of India, to
Bay of Bengal; approx. 1550
mi. long

The Holy River
The Sacred River

GANGRA, TURKEY. Ancient
town on tributary of Kizil
Irmak R. about 60 mi. N.E.
of Ankara

The City of the Synod

Gap in the Alps, The. See:
BRENNER PASS

Garden, America's Great Winter. See: IMPERIAL VAL-
LEY

Garden and Granary of Spain,
The. See: ANDALUSIA,
SPAIN

Garden City of Queensland, The.
See: TOOWOOMBA, AUS-
TRALIA

Garden Island, The. See: KAUAI

Garden of England, The. See:
HEREFORD, ENGLAND;
KENT, ENGLAND; WIGHT,
ISLE OF; WORCESTER-
SHIRE, ENGLAND

Garden of Erin, The. See:
CARLOW, IRELAND

Garden of Europe, The. See:
BELGIUM; ITALY

Garden of France, The. See:
INDRE-ET-LOIRE; TOURAINE

Garden of Helvetia, The. See:
THURGAU

Garden of India, The. See:
OUDH, INDIA

Garden of Italy, The. See:
SICILY

Garden of Scotland, The. See:
MORAYSHIRE

Garden of Spain, The. See:
ANDALUSIA, SPAIN

Garden of Sweden, The. See:
BLEKINGE, SWEDEN

Garden of the Argentine, The.
See: TUCUMAN

Garden of the World, The. See:
UNITED STATES OF AMER-
ICA

Gardens, The Town of Floating.
See: XOCHIMILCO, MEXICO

Gardens of the Sun, The. See:
MALAY ARCHIPELAGO

Gate, Golden. See: Golden
Gate, The

Gate of Asia, The. See: KA-
ZAN, U.S.S.R.

Gate of France, The Iron. See:
LONGWY, FRANCE

Gate of Italy, The. Gorge in
valley of Adige, Italy, near
Trento

Gate of Tears, The. See:

BAB EL MANDEB

Gate of the Inlet, The. See:
TOKYO, JAPAN

Gate of the Mediterranean, The.
See: GIBRALTAR, STRAIT
OF

Gate of the Talisman, The.
Portal in city of Baghdad,
Iraq, sit. on both sides of
Tigris R.

Gate to Mongolia, The. See:
WANCHUAN, MONGOLIA

Gated, The Hundred. See:
HECATOMPYLOS

Gated City, The Hundred. See:
THEBES, EGYPT

Gates, The Caspian. See:
DERBENT, RUSSIA IN
EUROPE

Gates, Hot. See: THERMO-
PYLAE

Gates, The Iron. See: DER-
BENT, RUSSIA IN EUROPE.
Iron Gates, The

Gates of Hercules, The. See:
Pillars of Hercules, The

Gates of Somnath, The Home
of the. See: SOMNATH,
W. INDIAN UNION

Gateway, The Derbent. See:
DERBENT, RUSSIA IN
EUROPE

Gateway to Central Asia, The.
See: KHYBER PASS

Gateway to India, The. See:
KHYBER PASS

Gateway to Nova Scotia, The
Inland. See: AMHERST,
NOVA SCOTIA

Gateway to Shensi, The. See:
TUNGKWAN, CHINA

GAZA, PALESTINE. City and
seaport of Egyptian-occupied
Palestine sit. about 3 mi.
from Mediterranean Sea

The City of Dagon

GELDERLAND, E. NETHER-
LANDS. Province of E.
Netherlands

The Dutch Paradise

Gem of the Desert, The. See:
GRAAF REINET, SOUTH
AFRICA

GENEVA, SWITZERLAND.
City of Geneva canton,
S.W. Switzerland, at S.
tip of Lake of Geneva on
Rhone R.

The City of Lost Causes
The City of Peace
The Intellectual Center of
Protestant Europe
The Rome of Protestantism

Gennesaret, The Lake of. See:
GALILEE, SEA OF

GENOA, ITALY. Seaport,
Genova province, Liguria,
N.W. Italy, 71 mi. S.S.W.
of Milan

The Birthplace of Columbus
The City of Palaces
The Superb

Gentle, The. See: FLORENCE,
ITALY

George, The City of St. See:
LOD, PALESTINE

GEORGE, LAKE. Lake of N.W.
New York State, in foothills
of Adirondack Mts., cover-

ing an area of approx. 44
sq. mi.

Lake Horicon

George, The Minature Lake.
See: GREENWOOD LAKE

George Town. See: PENANG,
MALAYSIA

German Athens, The. See:
WEIMAR, GERMANY

German Florence, The. See:
DRESDEN, GERMANY

GERMANY. Country of central
Europe, 182,426 sq. mi.,
sit. on Baltic Sea and North
Sea

The Fatherland
The Land of the Boar

Germany, The Armory of.
See: SUHL, GERMANY

Germany, The Eden of. See:
BADEN, AUSTRIA

Germany, The Sheffield of.
See: SOLINGEN, GERMANY

Germany's Waterloo. See:
BASTOGNE, BELGIUM

Gertie. See: Galloping Gertie

GETHSEMANE, PALESTINE.
In Biblical times small olive
grove sit. on Mount of
Olives about 3/4 mi. from
Jerusalem, where Jesus
withdrew with his disciples
on eve of Crucifixion

The Grotto of Agony

Gettysburg of the Pacific, The.
See: TRUK

GHEEL, BELGIUM. Commune,

Antwerp province, N. Bel-
gium; colony where model
system for dealing with the
insane is applied

The City of the Simple

GHENT, BELGIUM. City of
N. W. central Belgium sit.
at confluence of Schelde and
Lys Rivers

The Manchester of Belgium

Giant, The. See: NILE RIV-
ER

Giant, The Northern. See:
RUSSIA

Giant, The Sleeping. See:
CHINA

Giants, The Land of. See:
PACIFIC NORTHWEST

Giant's Causeway, The. Group
of basaltic columns on
coast of Antrim, N. Ireland

Giant's Grace, The. Height on
Adriatic shore of Bosporus
Strait, Turkey

Giant's Ladder, The. See:
MOFFAT TUNNEL

Gib, Old. See: GIBRALTAR

Gibbet Island. See: ELLIS
ISLAND

GIBEON, PALESTINE. Bib-
lical city in territory of
Benjamin, corresponding to
modern village of Eljib, 6
mi. N. W. of Jerusalem

Hill City

GIBRALTAR. British crown
colony sit. at southernmost
point of Iberian Peninsula

at junction of Mediterranean
Sea and Atlantic Ocean

Britannia's Lion
The Hill of Tarik
The Hub of the World
The Key of the Mediterran-
ean
Old Gib
The Rock
The Rock of Gibraltar
A Symbol of Modern Might

Gibraltar, The Rock of. See:
GIBRALTAR

GIBRALTAR, STRAIT OF.
Strait connecting Mediterranean Sea and Atlantic
Ocean sit. at southern tip
of Iberian Peninsula

The Gate of the Mediter-
ranean
The Straits

Gibraltar of America, The.
See: QUEBEC, CANADA

Gibraltar of North America,
The. See: SAINT JOHN'S,
NEWFOUNDLAND

Gibraltar of Scotland, The.
See: DUMBARTON, ROCK
OF

Gibraltar of the East, The.
See: ADEN, S.W. ARABIA;
CORREGIDOR ISLAND

Gibraltar of the New World, The.
See: DIAMOND, CAPE

Gibraltar of the Pacific, The.
See: PEARL HARBOR

Gibraltar of the South, The.
See: SIMONS-TOWN, SOUTH
AFRICA

GILEAD. See: RAMOTH GIL-
EAD, PALESTINE

GILES'S CHURCH.   See:
SAINT GILES'S CHURCH

Gimme Gimme Land.   See:
LIBYA (LIBIA)

Ginger, The Land of Green.
See: HULL, ENGLAND

Girdle of China, The.   See:
YANGTZE KIANG RIVER

Girt Isle, The Sea.   See:
ENGLAND

GIZA, EGYPT.   City of Egypt
sit. on W. bank of Nile R.
3 mi. S.W. of Cairo

The City of Pyramids

GLASGOW, SCOTLAND.   City
of S. central Scotland sit.
on both banks of Clyde R.

The City of the West
The Venice of the West

Glory of the East, The.   See:
PERSEPOLIS, PERSIA

Goat Island.   See: YERBA
BUENA

GOBI DESERT.   Desert of
central Asia, mostly in
Mongolia; about 500,000
sq. mi.

The Sandy Waste

God, The City of the Apostle
of.   See: MEDINA, SAUDI
ARABIA

God, The House of.   See:
BETHEL, PALESTINE

Goddess, The Jealous.   See:
EVEREST, MOUNT

Goddess-Mother, The.   See:
EVEREST, MOUNT

Goddess-Mother of the World,
The.   See: EVEREST,
MOUNT

Goddess Mountain, The.   See:
EVEREST, MOUNT

Godiva, The Home of Lady.
See: COVENTRY, ENG-
LAND

Godiva's Ride, The Town of.
See: COVENTRY, ENGLAND

Gods, The Abode of the.   See:
IDA MOUNTAINS

God's House.   See: LHASA
(LASSA), TIBET

GOLCONDA, INDIA.   Ruined
town of Hyderabad state,
S. central India, 5 mi. W.
of Hyderabad City

The City of Diamonds

Gold.   See: Field of the Cloth
of Gold, The

Gold, The Camel with the Feet
of.   See: LIBYA (LIBIA)

Gold, The City of Alluvial.
See: BENDIGO, AUSTRALIA

Gold, The Exactress of.   See:
BABYLON

Gold, The Path of.   See: MAR-
KET STREET, SAN FRAN-
CISCO, CALIFORNIA

Gold, The River of.   See:
AMERICAN RIVER

Gold Mine of Europe, The.
See: TRANSYLVANIA

Gold Purse of Spain, The.   See:
ANDALUSIA, SPAIN

Gold Tunnel, The Great.   See:

Golden Chersonese, The. See: MALAY PENINSULA (MALAYSIA)

Golden City, The. See: JOHANNESBURG, SOUTH AFRICA

Golden Fall, The. See: GULLFOSS

Golden Gate, The. Gate in Wall of Theodosius, Istanbul, Turkey; now walled up

Golden Horn, The. See: BOSPORUS

Golden Horn, The City on the. See: ISTANBUL, TURKEY

Golden House, The. Palace at Rome, Italy, built by Emperor Nero following fire of A.D. 64

Golden Lilies. See: Tank of the Golden Lilies, The

Golden Sand River, The. See: KINSHA

Golden Temple, The City of the. See: AMRITSAR, INDIA

Golden Triangle, The. Area of Burma, Laos and Thailand where opium poppies grown. Also main business section of Pittsburgh, Pa.

Golden Valley, The. E. part of Limerick, S.W. Ireland, Munster province, sit. on Shannon R.

Golden Weddings. See: House of Golden Weddings, The

Golfer's Paradise, The. See:

GOMORRAH. Ancient city in the Plain of Jordan; with Sodom, one of the Cities of the Plain

GOOD HOPE, CAPE OF. Headland on S.W. coast of Cape of Good Hope province, Union of S. Africa about 30 mi. from Capetown

The Cape
The Cape of Storms
The Extremity of the Continent
The Head of Africa
The Lion of the Sea
The Stormy Cape

GOODWIN SANDS. Dangerous shoals in Strait of Dover 7 mi. E. of town of Deal, England, on English Channel

The Ship-Swallower

Gordian Knot, The City of the. See: GORDIUM, ASIA MINOR

GORDIUM, ASIA MINOR. Ancient city of Phrygia on right bank of Sakaraya R., 50 mi. W.S.W. of Ankara

The City of the Gordian Knot

GOTHAM, ENGLAND. Village of Nottinghamshire, England, sit. 7 mi. S.W. of city of Nottingham

The Home of the Wise Men

GOTLAND. Island in Baltic Sea off S.E. coast of Sweden

The Eye of the Baltic

Government, The Home of
the People's. See: TERI-
OKI, U.S.S.R.

GOWRIE CHURCH. House of
worship located at Baffshire,
Scotland

The Kirk of Skulls

GRAAF REINET, SOUTH AF-
RICA. Town of Cape Col-
ony, S. Africa, sit. on
Sunday R., 135 mi. N.N.W.
of Port Elizabeth

The Gem of the Desert

Graham McNamee Mountains,
The. See: SIERRA MAD-
RE

Grahame's Dike. Roman wall
between Clyde and Forth
R.'s, Scotland

GRAHAMSTOWN, SOUTH AF-
RICA. Residential town of
S. Africa 75 mi. E.N.E.
of Port Elizabeth

The Most English Town in
South Africa

Grain Coast, The. See: LI-
BERIA

GRANADA, SPAIN. City of
Granada province, S. Spain,
80 mi. S.E. of Córdoba

The Last Moorish Strong-
hold in Spain

Granary of Athens, The. See:
KERCH, U.S.S.R.

Granary of California, The.
See: SAN JOAQUIN VAL-
LEY

Granary of Europe, The.
See: SICILY

Granary of Italy, The. See:
SICILY

Granary of Portugal, The.
See: ALENTEJO, PORTU-
GAL

Granary of Sicily, The. See:
CATANIA, SICILY

Granary of Spain, The. See:
ANDALUSIA, SPAIN

Granary of Spain, The Garden
and. See: ANDALUSIA,
SPAIN

Granary of the Roman World,
The. See: EGYPT

GRAND BANK. Shoals or
banks in Atlantic Ocean E.
and S. of Newfoundland

The Home of the Codfish

GRAND CANAL. Inland water-
way of N.E. China, reach-
ing from Tientsin to Hangchow, about 1000 mi. long

The Imperial River
The Transport River

Grand Canyon of American
Business, The. See:
BROADWAY, NEW YORK
CITY

Grand Canyon of the Pacific,
The. See: WAIMEA CAN-
YON

Grand Capital, The. See:
ANGKOR, CAMBODIA

GRAND PRE, NOVA SCOTIA.
Village of King's County,
W. Nova Scotia, 15 mi.
N.W. of Windsor, Ontario

The Great Meadow
The Home of Evangeline

GRANDE. See: RIO GRANDE

GRANDE, LA. See: CHAR-
TREUSE, LA GRANDE

Granite City, The. See: ABER-
DEEN, SCOTLAND

Grasshopper Hill. See: CHA-
PULTEPEC, MEXICO

Grave. See: Giant's Grave,
The

Gray Azores, The. See:
AZORES

Great American Desert, The.
Former name of semiarid
region between Sierra Ne-
vada and Rocky Mountains
of W. United States

Great Bed, The Home of the.
See: WARE, ENGLAND

Great Bridge, The. See:
BROOKLYN BRIDGE

GREAT BRITAIN. Largest
island in Europe and kingdom coextensive with the
island, comprising England,
Scotland and Wales

Albion
The Dumping Ground of
Europe
The Mistress of the Seas
The Old Country

GREAT BRITAIN. See also:
ENGLAND

Great Capital, The. See:
KHANBALIK, CHINA

Great Commercial Highway of
S.E. China, The. See:
SI-KIANG

Great Divide, The. See:
CONTINENTAL DIVIDE

Great Gold Tunnel, The. See:
SUTRO TUNNEL

Great Khan, The City of the.
See: KHANBALIK, CHINA;
PEKING, CHINA

Great King, The City of the.
See: JERUSALEM, PAL-
ESTINE

Great Lake, The. See: TAI-
HU; TONLE SAP

Great Lakes, The. Chain of
five fresh-water lakes in
central North America, consisting of: LAKE ERIE;
LAKE HURON; LAKE MICH-
IGAN; LAKE ONTARIO;
LAKE SUPERIOR

Great Meadow, The. See:
GRAND PRE, NOVA SCOTIA

Great Ox-Bow Route, The.
Butterfield overland mail
route, St. Louis, Mo., to
San Francisco, Calif., established by Congress in
1857

Great Plains, The. Continental
slope of central N. America
extending E. from Rocky
Mts. to margin of central
plains in U.S. and to margin of Laurentian Highlands
in Canada

Great River, The. See: GUA-
DALQUIVIR RIVER; MAHA-
NADI RIVER; MISSISSIPPI
RIVER; NILE RIVER

Great Rock of Italy, The. See:
CORNO, MONTE

Great Salt Desert, The. See:
DASHT-I-KAVIR

Great Sandy Island, The. See:
FRASER ISLAND

Great Sea, The. See: MED-
ITERRANEAN SEA

Great Smoke, The. See: LON-
DON, ENGLAND

Great Smokies, The. See:
GREAT SMOKY MOUNTAINS

GREAT SMOKY MOUNTAINS.
Range of Appalachian Mts.
extending along Tennessee/
N. Carolina border

The Great Smokies
The Playground of Two Na-
tions
The Rooftop of Eastern Amer-
ica
The Smokies

Great Stone Face, The. See:
CANNON MOUNTAIN

Great Street, That. See:
STATE STREET, CHICAGO,
ILL.

Great Temple, The Place of the.
See: KARNAK, UNITED
ARAB REPUBLIC

Great Thirst Land, The. See:
SOUTH AFRICA

Great Volga, The. See: VOL-
GA RIVER

GREAT WALL, THE. Defen-
sive wall extending 1250 mi. between Mongolia and China proper, built in 3rd century B. C.
by Emperor Shih Huang Ti

The Chinese Wall
The Great Wall of China
The Long Rampart

Great White Way, The. See:
BROADWAY, NEW YORK
CITY

Great Winter Garden, America's. See: IMPERIAL

VALLEY

Greater Britain. Collective
term for colonial dependen-
cies of British empire

Greece, The Eye of. See:
ATHENS, GREECE

Greece, The Light of. See:
CORINTH, GREECE

GREEN. See: GRETNA
GREEN, SCOTLAND

Green Ginger, The Land of.
See: HULL, ENGLAND

Green Isle, The. See: IRE-
LAND (EIRE)

Green of Ireland, the Gretna.
See: PORTPATRICK, SCOT-
LAND

Green Sea, The. See: PER-
SIAN GULF

GREENWICH VILLAGE. Resi-
dential section of borough of
Manhattan, New York City,
bounded by W. 14th St.,
Broadway, W. Houston St.
and West St.

The Center of Bohemia
The Hotbed of Bohemianism

GREENWOOD LAKE. Lake
sit. in New York and New
Jersey 40 mi. S. E. of New
York City

The Miniature Lake George

Gretna Green of Ireland, The.
See: PORTPATRICK, SCOT-
LAND

GRETNA GREEN, SCOTLAND.
Village of civil parish of
Gretna, Dumfries county,
Scotland, 9 mi. N. N. W. of
Carlisle

The Town of Many Marriages

GRISWOLD STREET, DETROIT, MICH. Street in business district of Detroit, Mich.

The Wall Street of Detroit

GROAT'S HOUSE. See: JOHN O' GROAT'S HOUSE

GROSS. See: SCHRECKHORN, GROSS

Grotto of Agony, The. See: GETHSEMANE, PALESTINE

Ground of Europe, The Dumping. See: GREAT BRITAIN

GRYNEION, ASIA MINOR. Ancient town on N.W. coast of Asia Minor, near Cyme

The City of Apollo

GUADALAJARA, SPAIN. Capital of Spanish province of same name sit. on Henares R., 35 mi. E.N.E. of Madrid

The Valley of Stones

GUADALQUIVIR RIVER. R. of S. Spain flowing from province of Jaén to Gulf of Cádiz; about 374 mi. long

The Great River

GUAIRA. Cataract in Paharà R. on boundary between Brazil and Paraguay

The Seven Falls

GUATEMALA. Republic of Central America, 42,044 sq. mi., sit. on Pacific Ocean

The Land of the Quetzal

GUERNSEY. One of the Channel Islands sit. in English Channel; approx. 25 sq. mi.

The Holy Island
The Home of Guernsey Cattle

GULF. See: CALIFORNIA, GULF OF; PERSIAN GULF

Gulf, The Key of the. See: CUBA

GULLFOSS. Waterfall 150 ft. high in Hvitá R., S.W. Iceland, near Geysir

The Golden Fall

Guns, The Town of the Barisàl. See: BARISAL, PAKISTAN

-H-

HAITI. Republic occupying W. third of Hispaniola I., W. Indies; 10,850 sq. mi.

The Black Republic

HAITIEN. See: CAP-HAITIEN, HAITI

HALEAKALA. Extinct volcano sit. on island of Maui, Hawaii; 10,025 ft. high

The House of the Sun

HALEMAUMAU. Fire pit of Kilauea crater on slope of Mauna Loa, volcano of Hawaii Island, Hawaii

The House of Everlasting Fire

Half a Continent. See: BRAZIL

Half Acre. See: Hell's

Half Acre

Half the World, The Banker
to. See: HONG KONG,
CHINA

HALL. See: FANEUIL HALL;
INDEPENDENCE HALL

Hall, The Home of the Cloth.
See: IEPER (YPRES),
BELGIUM

Hall Putsch, The Town of the
Beer. See: MUNICH, BA-
VARIA

HAMELN, GERMANY. Town
of Lower Saxony, W. Ger-
many, sit. at confluence of
Hamel and Weser R.'s, 25
mi. S.W. of Hanover

The Town of the Pied Piper

Hamlet, The Home of. See:
HELSINGÖR (ELSINORE),
DENMARK

Hamlet, The Town of. See:
HELSINGÖR (ELSINORE),
DENMARK

HAMMERFEST, NORWAY.
Town on island of Kvalöy
in county of Finnmark, Norway; the northernmost
town in Europe

The Farthest North

Han Cities, The. Wuhan. Collective names of Chinese
cities of Hankow, Hanyang
and Wuchang, sit. at Con-
fluence of Han and Yangtze
Kiang R.'s

Han River, The Mouth of the.
See: HANKOW, CHINA

Hangings, The Hill of the.
See: TYBURN HILL,

ENGLAND

HANKOW, CHINA. City of
S.E. Hupeh province, E.
central China, sit. on N.
bank of Yangtze Kiang R.,
E. of the Han R. opposite
Wuchang

The Mouth of the Han Riv-
er
The Pittsburgh of China

Hanoi Hilton, The. French
prison used as prisoner of
war camp in Hanoi, North
Viet Nam

Hanse, The Queen of the. See:
LÜBECK, GERMANY

Happy Islands, The. See:
CANARY ISLANDS

Happy Valley, The. See:
KASHMIR (CASHMERE);
TENNESSEE VALLEY

HARBOR. See: PEARL HAR-
BOR

Harbor, Sailors' Snug. See:
STATEN ISLAND

HARBOR BRIDGE. See: SYD-
NEY HARBOR BRIDGE

HARLEM. Former village,
now residential and busi-
ness district in borough of
Manhattan, New York
City

The Capital of the Negro
Population of the United
States

HARZ MOUNTAINS. Mountain
group in central Germany
between Elbe and Weser
R.'s S. of Brunswick

The Stronghold of Paganism

HAWAII. Island of Hawaiian
Island group sit. in N. cen-
tral Pacific Ocean 2090 mi.
W.S.W. of San Francisco

The Big Island

HAWAIIAN ISLANDS. Chain of
20 volcanic and coral islands
in N. central Pacific Ocean
2090 mi. W.S.W. of San
Francisco

Paradise
The Paradise of the Pacific

Head-Hunters, The Home of
the. See: NAGA HILLS

Head, The Bird's. See: VO-
GELKOP

Head of Africa, The. See:
GOOD HOPE, CAPE OF

HEADS. See: NOOSA HEADS

Headstream of the Nile, The
Most Remote. See: KA-
GERA

Health, The Land of. See:
BADAJOZ, SPAIN

Heart of England, The. See:
WARWICKSHIRE

Heart of Ireland, The. See:
ATHLONE, IRELAND

Heart of the Empire, The. See:
MOSCOW, RUSSIA

HECATOMPYLOS. Ancient city
of kingdom of Parthia sit.
at foot of S. slope of E. El-
burz mountains

The Hundred-Gated

Heel of England, The Achilles'.
See: IRELAND (EIRE)

Heel of Italy, The. See: SAL-

ENTINE PENINSULA

HELICE, GREECE. Ancient
city of Achaea, N. Pelop-
onnesus, S. Greece sit.
on N. part of shore of Gulf
of Corinth

The City of Poseidon

HELICON. Mountain in S.W.
Attica and Boetia dept., E.
central Greece near Gulf of
Corinth; 5738 ft. high

The Abode of Apollo and
the Muses

HELIOPOLIS, EGYPT. Holy
city of ancient Egypt sit.
about 6 mi. N.E. of Cairo;
center of sun worship during pre-Moslem Egyptian
civilization

The City of Ra
The City of the Sun

Hell. See also: Limbo

Hell, The Legendary Mouth of.
See: AVERNUS, LAKE

Hell, The Sailors'. See:
SHANGHAI, CHINA

Hell on Earth. See: ANDER-
SONVILLE PRISON

Hell on Wheels. Appellation of
newly formed American railroad towns during the mid-
19th century

Hell's Entrance. See: AVER-
NUS, LAKE

Hell's Half Acre. Portion of
battlefield at Spottsylvania,
Virginia, during American
Civil War (May, 1864)

Hell's Kitchen. Slum district
of New York City

HELSINGÖR (ELSINORE),
DENMARK. Seaport of
Frederiksborg county, N.
Sjaelland Island, Denmark

The Home of Hamlet
The Town of Hamlet
The Wapping of Denmark

Helvetia, The Garden of. See:
THURGAU

Helvetian Mountains, The. See:
ALPS, SWISS

Helvetian Republic, The. See:
SWITZERLAND

Hemisphere, The Switzerland
of This. See: PANAMA

Henry II. See: House of Henry II, The

HERACLEA, ITALY. Ancient
city of Luciana, Italy, near
Gulf of Taranto

The City of Pyrrhic Victory

HERAT, AFGHANISTAN.
City of Herat province,
N.W. Afghanistan, sit. on
Hari Rud R.

The Key of India

Hercules, The Gates of. See:
Pillars of Hercules, The

Hercules, The Pillars of. See:
Pillars of Hercules, The

HEREFORD, ENGLAND. Mu-
nicipal borough of W. England, sit. on Wye R., 47
mi. S.W. of Birmingham

The Garden of England

Hermit Nation, The. See:
KOREA

Hermitage, The. Home of

Andrew Jackson sit. near
Nashville, Tennessee

Hermits, The Place of the.
See: EINSIEDELN, SWITZ-
ERLAND

HERMON, MOUNT. Mountain
of the Anti-Liban range,
S.W. Levant States, 9232
ft. high

Old Man Mountain
Snow Mountain

Hero of Africa, The. See:
CUBA

Herod, The Once-Royal City
of. See: TIBERIAS, PAL-
ESTINE

Heroin Crossroads of South
America, The. See: PAR-
AGUAY

Herophile, The Home of. See:
ERYTHRAE, LYDIA

Herring Pond, The. See: AT-
LANTIC OCEAN

Hibernia. See: IRELAND
(EIRE)

Hidden Island. See: CEPHA-
LONIA

High Capital, The Sky. See:
LA PAZ, BOLIVIA

Highlands. Designation of
mountainous portion of Scotland extending N. and W. of
the Grampians

Highlands, The Key to the.
See: STIRLING, SCOTLAND

HIGHWAY. See: C.C.C. HIGH-
WAY, THE

Highway. See: Inter-American
Highway

Highway, The Three C's. See: C.C.C. HIGHWAY, THE

Highway of Southeast China, The Great Commercial. See: SI-KIANG

Highway to India, The. See: SUEZ CANAL

HILL. See: BROKEN HILL, AFRICA; CAPITOL HILL; TEMPLE HILL; TYBURN HILL, ENGLAND

Hill. See: Gallows Hill; Hungry Hill; Sagamore Hill

Hill, The. See: CAPITOL HILL

Hill, Grasshopper. See: CHAPULTEPEC, MEXICO

Hill, The Holy. See: ATHOS, MOUNT

Hill, Salt Pan. See: ZOUTPANSBERG

Hill City. See: GIBEON, PALESTINE

Hill City, The. See: Rome, Italy

Hill of Tarik, The. See: GIBRALTAR

Hill of the Hangings, The. See: TYBURN HILL, ENGLAND

Hilled City, The Seven. See: ROME, ITALY

HILLS. See: NAGA HILLS

Hills. See: Black Hills, The

Hills, The City of Seven. See: ROME, ITALY

Hills, The Crystal. See: WHITE MOUNTAINS

Hills, The River of Steep. See: HUDSON RIVER

Hills, Silver. See: TEGUCIGALPA, HONDURAS

Hilton. See: Hanoi Hilton, The

HIMALAYA, THE. Mountain system of S. central Asia lying between N. India and Tibet containing the highest peaks in the world

The Abode of Snow

Hindus, The Place of the. See: HINDUSTAN

HINDUSTAN. Indian peninsula N. of Deccan R., lying between Himalaya and Vindhya mountains

The Land of the Sun
The Place of the Hindus

Hindustan, The Rome of. See: AGRA, INDIA

His Holiness, The Home of. See: VATICAN CITY

HISPANIOLA. Island of W. Indies lying E. of Cuba and W. of Puerto Rico in N. central Caribbean Sea

Little Spain (La Española)

Hitler's Waterloo. See: BASTOGNE, GERMANY

HO. See: HWANG HO; PEI-HO

Hogen-Mogen. See: NETHERLANDS, THE (HOLLAND)

HOKKAIDO, JAPAN. One of the

principal and northernmost of islands of Japanese archipelago

The Circuit of the Northern Seas

Hole, The City of the Black. See: CALCUTTA, INDIA

Hole of Calcutta. See: Black Hole of Calcutta, The

Holiness, The Home of His. See: VATICAN CITY

HOLLAND. See: NETHERLANDS, THE (HOLLAND)

Holland, New. See: AUSTRALIA

Holy City, The. See: ALLAHABAD, INDIA; BENARES, INDIA; CUZCO, PERU; DAMASCUS, SYRIA; JERUSALEM, PALESTINE; KIEV, U.S.S.R.; MOSCOW, RUSSIA; ROME, ITALY

Holy City of Islam, The. See: MECCA, SAUDI ARABIA; MEDINA, SAUDI ARABIA

Holy Empire, The. See: HOLY ROMAN EMPIRE

Holy Hill, The. See: ATHOS, MOUNT

Holy Island, The. See: GUERNSEY; IRELAND (EIRE); LINDISFARNE; RÜGEN, GERMANY

Holy Land, The. See: ARABIA; ELIS; INDIA; PALESTINE

Holy Mother of the Russians, The. See: MOSCOW, RUSSIA

Holy Mountain, The. See:

ATHOS, MOUNT

Holy Rapids, The. See: PYHÄKOSKI

Holy River, The. See: GANGES RIVER

HOLY ROMAN EMPIRE. Designation applied to an amorphous political entity of W. Europe originated by Pope Leo III in A.D. 800 and in nominal existence until 1806

The Empire of the West
The Holy Empire
The Roman Empire

HOLYWELL, WALES. Urban district of Flintshire, N.E. Wales; site of St. Winifred's Well

The Lourdes of Wales

Home, Medea's. See: COLCHIS

Home, The Pirate's. See: BARATARIA BAY

Home, Sweet Home House. Boyhood home of John Howard Payne (1791-1852), American author and actor, at Easthampton, Long Island, N.Y.

Home Counties, The. Collective designation of English counties of Essex, Hertford, Kent, Middlesex and Surrey

Home of Armagnac Brandy, The. See: ARMAGNAC, FRANCE

Home of Banbury Cakes, The. BANBURY, ENGLAND

Home of Bock Beer, The. See: EINBECK, WEST GERMANY

Home of Brandy, The. See:
CONDOM, FRANCE

Home of Brie Cheese, The.
See: BRIE, FRANCE

Home of Chartreuse Liqueur,
The. See: CHARTREUSE,
LA GRANDE

Home of Cheese, The. See:
LIMBURG, BELGIUM

Home of Chu Hsi, The. See:
LUNGKI, CHINA

Home of David, The. See:
BETHLEHEM, PALESTINE

Home of Dresden China, The.
See: DRESDEN, GERMANY

Home of Early Chinese Agriculture, The. See: SHAN-
SI, CHINA

Home of Eli and Samuel, The.
See: SHILOH, PALESTINE

Home of Epsom Salts, The.
See: EPSOM, ENGLAND

Home of Evangeline, The. See:
GRAND PRE, NOVA SCOTIA

Home of Fighting for Fun, The.
See: DONNYBROOK, IRE-
LAND

Home of Guernsey Cattle, The.
See: GUERNSEY

Home of Hamlet, The. See:
HELSINGÖR (ELSINORE),
DENMARK

Home of Herophile, The. See:
ERYTHRAE, LYDIA

Home of His Holiness, The.
See: VATICAN CITY

Home of Kidderminster Car-

pets, The. See: KIDDER-
MINSTER, ENGLAND

Home of Lady Godiva, The.
See: COVENTRY, ENG-
LAND

Home of Lazarus, The. See:
BETHANY, PALESTINE

Home of Madeira Wines, The.
See: MADEIRA

Home of Malmsey, The. See:
MONEMVASIA, GREECE

Home of Malvasia, The. See:
MONEMVASIA, GREECE

Home of Neckar Wine, The.
See: ESSLINGEN, WEST
GERMANY

Home of Odysseus, The. See:
ITHACA

Home of Parmesan Cheese,
The. See: PARMA, ITALY

Home of Pilsener Beer, The.
See: PLZEN (PILSEN)

Home of Port Wine, The. See:
OPORTO, PORTUGAL

Home of Robin Hood, The.
See: SHERWOOD FOREST

Home of St. Bernard, The.
See: CLAIRVAUX, FRANCE

Home of the Altar Stones, The.
See: STONEHENGE, ENG-
LAND

Home of the Arlberg Skiing
Technique, The. See: ARL-
BERG, AUSTRIA

Home of the Bard, The. See:
STRATFORD-UPON-AVON,
ENGLAND

Home of the Bloody Assizes,

The. See: DORCHESTER, ENGLAND

Home of the Book of Kells, The. See: KELLS, IRE-LAND

Home of the Brave, The Land of the Free and the. See: UNITED STATES OF AMER-ICA

Home of the Castlebar Races, The. See: CASTLEBAR, IRELAND

Home of the Caves, The. See: NOTTINGHAM, ENGLAND

Home of the Cloth Hall, The. See: IEPER (YPRES), BEL-GIUM

Home of the Clydesdale Horse, The. See: CLYDESDALE, SCOTLAND

Home of the Codfish, The. See: GRAND BANK

Home of the Coffee Plant, The. See: KAFA (KAFFA)

Home of the Constitutions, The. See: CLARENDON PARK, ENGLAND

Home of the Dards, The. See: HUNZA, INDIA

Home of the Derby, The. See: CHURCHILL DOWNS, LOUIS-VILLE, KY.; EPSOM DOWNS, ENGLAND

Home of the Dragon Lizard, The. See: KOMODO

Home of the Gates of Somnath, The. See: SOMNATH, W. INDIAN UNION

Home of the Great Bed, The.

See: WARE, ENGLAND

Home of the Head-Hunters, The. See: NAGA HILLS

Home of the Houdan Fowl, The. See: HOUDAN, FRANCE

Home of the Moonrakers, The. See: WILTSHIRE

Home of the North Magnetic Pole, The. See: BOOTHIA PENINSULA, CANADA

Home of the Oracle, The. See: DELPHI, GREECE

Home of the Passion Play, The. See: OBERAMMERGAU, BA-VARIA

Home of the People's Govern-ment, The. See: TERIOKI, U.S.S.R.

Home of the Peterborough Ca-noe, The. See: PETER-BOROUGH, CANADA

Home of the Pope, The. See: VATICAN CITY

Home of the Prisoner, The. See: CHILLON

Home of the Reformation, The. See: WITTENBERG, GER-MANY

Home of the Republic of China, The. See: TAIWAN (FOR-MOSA)

Home of the Rhodesian Man, The. See: BROKEN HILL, AFRICA

Home of the Rice Lake Canoe, The. See: PETERBOROUGH, CANADA

Home of the Sabines, The.

See: CURES, ITALY

Home of the Sibyl, The. See:
CUMAE, ITALY

Home of the Tailtean Games,
The. See: TELLTOWN,
IRELAND

Home of the Telescope, The.
See: SLOUGH, ENGLAND

Home of the Tortoise, The.
See: GALAPAGOS IS-
LANDS

Home of the Vampire, The.
See: TRANSYLVANIA

Home of the Wise Men, The.
See: GOTHAM, ENGLAND

Home of Theseus, The. See:
TROEZEN, GREECE

Home of Tobacco, The. See:
LATAKIA, SYRIA

Home of Turkish Carpets, The.
See: USAK, TURKEY

Home of White Wines, The.
See: CHABLIS, FRANCE

Home of William Tell, The.
See: ALTDORF, SWITZ-
ERLAND

Home of Wines, The. See:
BORDEAUX, FRANCE

Home of Worcestershire
Sauce, The. See: WOR-
CESTER, ENGLAND

Homes, A City of. See:
TORONTO, CANADA

HONG KONG, CHINA. Brit-
ish crown colony of S.E.
China sit. E. of mouth of
Pearl R. 90 mi. S. of
Canton

The Banker to Half the
World

HONITON, ENGLAND. Mu-
nicipal borough of Devon-
shire, S.W. England, 16
mi. N.E. of Exeter

The City of Lace

Honolulu, The Long Branch of.
See: WAIKIKI BEACH

Hood, The Home of Robin.
See: SHERWOOD FOREST

Hood's Forest, Robin. See:
SHERWOOD FOREST

HOPE. See: GOOD HOPE,
CAPE OF

Horicon, Lake. See: GEORGE,
LAKE

HORN, CAPE. Rock at S.
extremity of S. America
sit. on Horn Island

The Cape
The Horn

Horn, The. See: HORN,
CAPE

Horn, The City on the Golden.
See: ISTANBUL, TURKEY

Horn, The Golden. See: BOS-
PORUS

Horn of Africa, The. See:
SOMALILAND, AFRICA

Hornet's Nest, The. Battlefield area at American Civil War encounter of Shiloh,
Tenn., April 5/6, 1862

Horse, The Home of the Clydesdale. See: CLYDESDALE,
SCOTLAND

Horse Latitudes. Either of

two belts of latitudes located over the oceans at about 30° to 35° N. and S., characterized by high barometric pressure, calms and light variable winds

HÖRSEL BERGE, GERMANY. Mountains of N.W. Thuringia, Germany, sit. between Gotha and Eisenach

Venus's Court

Horus, The Town of. See: DAMANHUR, UNITED ARAB REPUBLIC

Hospitable Sea, The. See: BLACK SEA

Hot Gates. See: THERMO-PYLAE

Hotbed of Bohemianism, The. See: GREENWICH VILLAGE

HOUDAN, FRANCE. Village of Seine-et-Oise dept., N. France

The Home of the Houdan Fowl

HOUSE. See: JOHN O' GROAT'S HOUSE; WHITE HOUSE

House. See: Golden House, The; Home, Sweet Home House; Treason House

House, The Former Western White. See: SAN CLEM-ENTE, CALIFORNIA

House, God's. See: LHASA (LASSA), TIBET

House, The Western White. See: SAN CLEMENTE, CALIFORNIA

House of Bread, The. See:

BETHLEHEM, PALESTINE

House of Deity, The. See: BETHLEHEM, PALESTINE

House of Everlasting Fire, The. See: HALEMAUMAU

House of God, The. See: BETHEL, PALESTINE

House of Golden Weddings, The. House at Quincy, Mass., where golden weddings of various members of Adams family celebrated

House of Henry II, The. Sixteenth century palace and cathedral, now library and museum, sit. at La Rochelle, French seaport on Bay of Biscay, 124 mi. S.W. of Tours

House of Mercy, The. See: BETHESDA

House of the Sun, The. See: HALEAKALA

House with the Eagles, The. Home of Captain Churchill, American privateer, near Bristol, R.I.

HOWE, CAPE. Promontory of Australia, sit. on border of New S. Wales and Victoria; extreme S.E. point of Australia

The Farthest South

HU. See: SI-HU; TAI-HU

Hub of New York, The. See: COLUMBUS CIRCLE, NEW YORK CITY

Hub of the Orient, The. See: TAIPEI, FORMOSA

Hub of the Universe, The.

State House located at Boston, Mass.

Hub of the World, The. See:
GIBRALTAR

HUDSON RIVER. River of E.
New York State, flowing
from Adirondack Mts. into
upper New York Bay

The Rhine of America
The River of Steep Hills

Hudson River of the West, The.
See: MISSISSIPPI RIVER

Hugh Lloyd's Pulpit. Stone
formation in Merioneth-
shire, Wales

HULL, ENGLAND. County
borough of E. Riding, Yorkshire, N. England, at confluence of Humber and Hull
R.'s, 157 mi. N. of London

The Land of Green Ginger

Humble Submission, The Town
of. See: CANOSSA, ITALY

Hump, The. Ranges at E. end
of Himalaya mountain sys-
tem extending along border
between N. India and Tibet

Hundred-Gated, The. See:
HECATOMPYLOS

Hundred-Gated City, The. See:
THEBES, EGYPT

Hundred Thousand Islands, The.
See: LACCADIVE ISLANDS

Hundred Towers, The City of
a. See: PAVIA, ITALY

HUNGARY. See: AUSTRIA-
HUNGARY

Hungry Dragon, The. See:
CHINA

Hungry Hill. Campsite of
General Ambrose Burn-
side's Civil War troops
near Warrenton, Va. (1862)

Hunters, The Home of the
Head. See: NAGA HILLS

Hunters, The Seven. See:
FLANNAN ISLANDS

HUNZA, INDIA. District of
N. Kashmir, N. India

The Home of the Dards

Hurdles, The Town of the
Ford of the. See: DUB-
LIN, IRELAND

HURON, LAKE. See: Great
Lakes, The

Husain, The City of. See:
KERBELA, IRAQ

HWANG HO. River of N. China, second longest R. in
the country; approx. 2700
mi. long

China's Sorrow
The Sea of Stars
The Yellow River

Hyacinth and Ruby, The Land
of the. See: SRI LANKA
(CEYLON)

HYBLA, SICILY. Ancient
town of Sicily sit. on slope
of Mount Etna; famous for
its honey

The Empire of Bees
The Modern Paternò

-I-

Ice, The City Under the. See:
CAMP CENTURY, GREEN-
LAND

Ice Island, The. See: ANT-
   ARCTICA

Icebox of Siberia, The. See:
   OIMYAKON, RUSSIA

IDA MOUNTAINS. Mountain
   range of N. W. Asia Minor,
   S. E. of site of ancient
   Troy

   The Abode of the Gods

IEPER (YPRES), BELGIUM.
   Commune of W. Flanders
   province, N. W. Belgium

   The Home of the Cloth Hall
      Wipers

IGLESIAS, SARDINIA. Com-
   mune of Cagliari province,
   S. Sardinia, near W. coast,
   32 mi. W. N. W. of Cagliari

   The City of the Mines

ILLIMANI. Mountain in An-
   des range located in W.
   Bolivia 40 mi. S. E. of La
   Paz

   The Snow-Covered Mountain

Immortality, The City of the
   Pool of. See: AMRITSAR,
   INDIA

Imperial City, The. See:
   ROME, ITALY

Imperial River, The. See:
   GRAND CANAL

IMPERIAL VALLEY. Valley
   of Imperial County, California, sit. in S. E. corner
   of the state

   America's Great Winter
      Garden

INCHCAPE ROCK. Rock in

N. Sea about 11 mi. S. E.
   of Arbroath, Scotland, bear-
   ing lighthouse and covered
   by sea at high tide

   The Bell Rock

INDEPENDENCE HALL. Old
   State House at Philadelphia,
   Pa., scene of significant
   events in 18th-century American history

   The Birthplace of Liberty
   The Cradle of American
      Liberty

INDIA. Subcontinent and country of S. Asia, S. of Him-
   alayas between Arabian Sea
   and Bay of Bengal

   The Holy Land
   The Jewel of the British
      Empire
   The Land of the Veda
   The Lost Kingdom

India, The Garden of. See:
   OUDH, INDIA

India, The Gateway to. See:
   KHYBER PASS

India, The Highway to. See:
   SUEZ CANAL

India, The Islands of. See:
   MALAY ARCHIPELAGO

India, The Key of. See:
   HERAT, AFGHANISTAN

INDIAN STATES. Various
   semi-independent areas
   in India ruled by native
   Indians; subjected to British authority until August,
   1947, when made nomin-
   ally independent states

   The Native States

INDIANOLA PEAK. Mountain

of S. Brewster County, W.
Texas, 5240 ft. high

The Elephant Tusk

INDRE-ET-LOIRE. Depart-
ment of France in N.W.
central part of the country,
2377 sq. mi. in area

The Garden of France

Industrial Metropolis of the
Orient, The. See: OSAKA,
JAPAN

Industrious, The. See: LUC-
CA, ITALY

Infamy, The Wall of. See:
BERLIN WALL

Infant Mississippi River, The.
See: ITASCA, LAKE

Infant Mississippi River,
Nicollet's. See: ITASCA,
LAKE

Inhospitable Sea, The. See:
BLACK SEA

Inland Gateway to Nova Scotia,
The. See: AMHERST, NO-
VA SCOTIA

Inlet, The Gate of the. See:
TOKYO, JAPAN

Inn of the Kings, The. See:
PARIS, FRANCE

Innisfail. See: IRELAND
(EIRE)

Insula Sanctorum. See: IRE-
LAND (EIRE)

Intellectual Center of Protestant Europe, The. See:
GENEVA, SWITZERLAND

Intelligence, The City of.

See: BERLIN, GERMANY

Inter-American Highway. Name
given the section of the Pan-
American Highway extending
from Laredo, Texas, to
Panama City, Panama (approx.
3300 mi.)

INUTIL BAY. Large inlet on
N.W. coast of Tierra del
Fuego, Chile

Useless Bay

Ionia, The Crown of. See:
SMYRNA, ASIA MINOR

IRELAND (EIRE). Second
largest island, after Great
Britain, of British Isles,
sit. in N. Atlantic Ocean

The Achilles' Heel of Eng-
land
The Emerald Isle
The Emerald of Europe
Erin
The Green Isle
Hibernia
The Holy Island
The Island of Saints
The Isle of Saints
Innisfail
Insula Sanctorum
The Land of Destiny
The Land of the Potato
The Land of the Shamrock
The Owld Country
The Owld Sod
Potato Land
The Sacred Island
The Sister Isle
Spudland
Wolfland

Ireland. See: Russian Ireland

Ireland, The Athens of. See:
BELFAST, IRELAND; CORK,
IRELAND

Ireland, The Gretna Green of.

See: PORTPATRICK, SCOT-
LAND

Ireland, The Heart of. See:
ATHLONE, IRELAND

Ireland, The Orchard of. See:
ARMAGH, IRELAND

Iron Gate of France, The. See:
LONGWY, FRANCE

Iron Gates, The. Gorge in
Danube R. 2 mi. long sit.
on Yugoslav-Romanian bor-
der at W. extremity of
Transylvanian Alps

Iron Gates, The. See also:
DERBENT, RUSSIA IN
EUROPE

IRRAWADDY RIVER. River
of central Burma formed
by confluence of Mali and
Nmai R.'s flowing to Bay
of Bengal; about 1350 mi.
long

The Father of Waters

ISABELA, NORTH DOMINICAN
REPUBLIC. Cape and port
of N. Dominican Republic,
Hispaniola Island

Columbus's First Settlement

ISCHIA. Island in Bay of
Naples

The Smiling Island

ISHMONIE, EGYPT. Ruined
city of upper Egypt

The Petrified City

Islam, The Holy City of. See:
MECCA, SAUDI ARABIA;
MEDINA, SAUDI ARABIA

ISLAND. See: ALCATRAZ

ISLAND; AQUIDNECK IS-
LAND; BEDLOE'S ISLAND;
BLOCK ISLAND; CORREGI-
DOR ISLAND; ELLIS ISLAND;
FALSTER ISLAND; FRASER
ISLAND; JEKYLL ISLAND;
LIBERTY ISLAND; PITCAIRN
ISLAND; REST ISLAND,
MINN.; SINGAPORE ISLAND;
STATEN ISLAND; SUMBA IS-
LAND; WELFARE ISLAND

Island, The. See: ALCATRAZ
ISLAND; WELFARE ISLAND

Island, America's Devil's. See:
ALCATRAZ ISLAND

Island, The Big. See: HA-
WAII

Island, Bird. See: ALCA-
TRAZ ISLAND

Island, Blackwell's. See:
WELFARE ISLAND

Island, The "Bounty." See:
PITCAIRN ISLAND

Island, Crusoe's. See: MAS
A TIERRA

Island, Empress Josephine's.
See: MARTINIQUE

Island, The Garden. See:
KAUAI

Island, Gibbet. See: ELLIS
ISLAND

Island, Goat. See: YERBA
BUENA

Island, The Great Sandy. See:
FRASER ISLAND

Island, Hidden. See: CEPH-
ALONIA

Island, The Holy. See: GUERN-
SEY; IRELAND (EIRE); LIN-

DISFARNE; RÜGEN, GER-
MANY

Island, The Ice. See: ANT-
ARCTICA

Island, Liberty. See: BED-
LOE'S ISLAND

Island, Lost. See: CEPHA-
LONIA

Island, Oyster. See: ELLIS
ISLAND

Island, Pelican. See: ALCA-
TRAZ ISLAND

Island, The Ringing. See:
ENGLAND

Island, The Sacred. See:
IRELAND (EIRE)

Island, Sandalwood. See:
SUMBA ISLAND

Island, Selkirk's. See: MAS
A TIERRA

Island, The Smiling. See:
ISCHIA

Island, Sulfur. See: IWO
JIMA

Island, The Sunken. See:
ATLANTIS

Island, The Tight Little. See:
ENGLAND; TAIWAN (FOR-
MOSA)

Island, Tin Can. See: NIU-
AFOO

Island, The Vaunted. See:
SINGAPORE ISLAND

ISLAND, WATLINGS. See:
SAN SALVADOR (WAT-
LINGS ISLAND)

Island Aircraft Carriers,

The Stationary. See: MAR-
IANA ISLANDS

Island City, The. See: MON-
TREAL, CANADA

Island Continent, The. See:
AUSTRALIA

Island Nation, The. See:
SINGAPORE ISLAND

Island of Mutineers, The.
See: PITCAIRN ISLAND

Island of Pelicans, The. See:
ALCATRAZ ISLAND

Island of Promise, The. See:
PUERTO RICO

Island of Sappho, The. See:
LESBOS

Island of Stone Money, The.
See: YAP

Island of the Fountain of
Youth, The. See: BIMINI

Island of the Lepers, The.
See: MOLOKAI

Island of the Saints, The.
See: IRELAND (EIRE)

Island of the West, The Emerald. See: MONTSERRAT

Island Sea, The. See: ARAL,
SEA OF

ISLANDS. See: APOSTLE
ISLANDS; ARAN ISLANDS;
CANARY ISLANDS; DODE-
CANESE ISLANDS; FLAN-
NAN ISLANDS; GALAPAGOS
ISLANDS; HAWAIIAN IS-
LANDS; KURIL (KURILE)
ISLANDS; LACCADIVE IS-
LANDS; LIPARI ISLANDS;
MARIANA ISLANDS; PRIB-
LOFF ISLANDS; SAFETY
ISLANDS (ISLES DU SALUT);

SAMOA ISLANDS; SHET-
LAND ISLANDS; TONGA
ISLANDS

Islands, The Aeolian. See:
LIPARI ISLANDS

Islands, Devil's (Isles du Diable). See: SAFETY IS-
LANDS (ISLES DU SALUT)

Islands, The Empire of the.
See: MALAY ARCHIPEL-
AGO

Islands, The Flower of. See:
CUBA

Islands, The Friendly. See:
TONGA ISLANDS

Islands, The Fur Seal. See:
PRIBLOFF ISLANDS

Islands, The Happy. See:
CANARY ISLANDS

Islands, The Hundred Thousand.
See: LACCADIVE ISLANDS

Islands, Navigators. See:
SAMOA ISLANDS

Islands, The Other. See:
OVERIGE EILANDEN

Islands, The Smoking. See:
KURIL (KURILE) ISLANDS

Islands, The Spice. See:
MOLUCCAS

Islands, Twelve. See: DO-
DECANESE ISLANDS

Islands of Dogs, The. See:
CANARY ISLANDS

Islands of India, The. See:
MALAY ARCHIPELAGO

Islands of Ponies, The. See:
SHETLAND ISLANDS

Islands of the Blessed, The.
See: CANARY ISLANDS

ISLE. See: MAN, ISLE OF;
WIGHT, ISLE OF

Isle, The Emerald. See: IRE-
LAND (EIRE)

Isle, The Ever-Faithful. See:
CUBA

Isle, The Green. See: IRE-
LAND (EIRE)

Isle, The Lovely. See: COR-
FU

Isle, The Queen-Faithful. See:
CUBA

Isle, The Sea-Girt. See: ENG-
LAND

Isle, The Sister. See: IRE-
LAND (EIRE)

Isle, The Tight Little. See:
ENGLAND

Isle De Dabney, The. See:
MADEIRA

Isle of Capri, The. See:
CAPRI, ITALY

Isle of No Return, The. See:
ALCATRAZ ISLAND

Isle of Peace, The. See:
AQUIDNECK ISLAND

Isle of Saints, The. See:
IRELAND (EIRE)

Isle of the Leper's Well, The.
See: STRONSAY

Isle of the Tailless Cat, The.
See: MAN, ISLE OF

Isles, The Fortunate. See:
CANARY ISLANDS

Isles, The River of a Thousand. See: WISCONSIN RIVER

Isles du Diable (Devil's Islands). See: SAFETY ISLANDS (ISLES DU SALUT)

ISLES DU SALUT. See: SAFETY ISLANDS (ISLES DU SALUT)

Isles of Izu, The Seven. See: IZU SHICHITO

ISTANBUL, TURKEY. City of N.W. Turkey sit. on both sides of Bosporus R. between Black Sea and Sea of Marmara

The City on the Golden Horn

Italian Boot, The Toe of the. See: CALABRIA, ITALY

ITALY. Republic of S. Europe comprising the boot-shaped peninsula extending into Mediterranean Sea plus certain islands, including Sicily and Sardinia

The Garden of Europe
The Land of Plenty
The Land of Song
Lavinia
Saturnia

Italy. See: Gate of Italy, The

Italy, The Garden of. See: SICILY

Italy, The Granary of. See: SICILY

Italy, The Great Rock of. See: CORNO, MONTE

Italy, The Heel of. See: SALENTINE PENINSULA

ITASCA, LAKE. Lake of S.E. Clearwater County, N. Minn.; source of Mississippi R.

The Infant Mississippi River
Nicollet's Infant Mississippi River
The Source of the Mississippi

ITHACA. Island in Ionian Sea off W. coast of Greece and N.E. of Cephalonia

The Home of Odysseus

ITZA. See: CHICHEN ITZA

Itza, The Mouth of the Wells of. See: CHICHEN ITZA

IVANOVO, RUSSIA. City of Soviet Russia sit. in Ivanov Industrial Area 145 mi. N.E. of Moscow

The Soviet Manchester

IWO JIMA. Large volcanic island sit. in Pacific Ocean 710 mi. S. of Tokyo

Sulfur Island

IXTACIHUATL. Extinct volcano in Mexico sit. 40 mi. S.S.E. of Mexico City

The Fat Woman
The White Woman

IZALCO. Active volcano of El Salvador, sit. about 10 mi. N. of Sonsonate; creates glow visible from Pacific Ocean

The Lighthouse of El Salvador

IZU SHICHITO. Group of volcanic islands in Pacific Ocean off Izu Peninsula on

S. E. coast of Honshu, S.
of Yokohama, Japan

The Seven Isles of Izu

-J-

JABUD. See: MASJID JABUD

JAFFA, ISRAEL. Seaport of
Israel sit. on Mediterranean
Sea adjacent to Tel Aviv

The Pilgrim Port for Jerusalem

Jagannath, The City of. See:
PURI, INDIA

Jamesheed, The Throne of.
See: PERSEPOLIS, PERSIA

JAMI. See: MASJID JAMI

JANEIRO. See: RIO DE JA-
NEIRO, BRAZIL

JAPAN. Constitutional island
monarchy in Pacific Ocean
off coast of N. E. Asia

The Britain of the Far East
The Land of the Rising Sun
The Land of the Rising Yen

Japan, The Paris of. See:
KYOTO, JAPAN; OSAKA,
JAPAN

Japan, The Parnassus of.
See: FUJI (FUJIYAMA),
MOUNT

Japan, The Pittsburgh of. See:
YAWATA, JAPAN

Japan, The Venice of. See:
OSAKA, JAPAN

JAVA. Island in Malay Archipelago sit. in Greater Sun-

da Islands group S. E. of
Sumatra and S. of Borneo

The Queen of the Eastern
Archipelago

Java Trough, The. See:
WHARTON DEEP

Jealous Goddess, The. See:
EVEREST, MOUNT

JEBEL MUSA. Sacred moun-
tain sit. on peninsula of
Sinai between Gulf of Suez
and Gulf of Akaba where
Moses is said to have re-
ceived the Ten Command-
ments from Jehovah

Mount Sinai
The Mountain of Moses

Jebusites, The City of the.
See: JERUSALEM, PAL-
ESTINE

Jehoshaphat, The Valley of.
See: KIDRON WADI

JEKYLL ISLAND. Island in Atlantic Ocean off mainland of
Glynn County, S. E. Georgia

The Millionaire's Resort

JENA, GERMANY. City of
Thuringia, E. Germany,
56 mi. S. W. of Leipzig

Luther's Sanctuary

JERUSALEM, PALESTINE.
Partitioned city of the Palestine region in Asia, 35 mi. E.
of the Mediterranean Sea

The City of David
The City of Peace
The City of the Great King
The City of the Jebusites
The Holy City
The Sacred City

Jerusalem, The Pilgrim Port
for. See: JAFFA, ISRAEL

Jewel, The Tarnished. See:
ALCATRAZ ISLAND

Jewel of the British Empire,
The. See: INDIA

JIMA. See: IWO JIMA

JOAQUIN. See: SAN JOA-
QUIN VALLEY

JOHANNESBURG, SOUTH AF-
RICA. City of S. Transvaal,
N.E. Union of S. Africa, 300
mi. N.W. of Durban

The Golden City

JOHN O' GROAT'S HOUSE.
Point on N. coast of
Caithness, N. Scotland;
usual terminus of races
the length of Great Brit-
ain from Land's End

The Finish Line

John, The Land of Prester.
See: ETHIOPIA (ABYS-
SINIA)

JOHN'S. See: SAINT JOHN'S,
NEWFOUNDLAND

JOHN'S RIVER. See: SAINT
JOHN'S RIVER

John's Trail. See: CHIS-
HOLM TRAIL

JÖNKÖPING, SWEDEN. Cap-
ital of Swedish county of
same name, 230 mi. S.W.
of Stockholm

The City of Matches

JORDAN RIVER. River of
Palestine flowing from
Anti-Liban Mountains W.

of Mt. Hermon to Dead
Sea; 45 mi. long

The Descender

JOSAS. See: JOUY-EN-
JOSAS, FRANCE

Josephine's Island, Empress.
See: MARTINIQUE

JOUY-EN-JOSAS, FRANCE.
Commune of Sainte-et-Oise
dept., N. France, S.E. of
Versailles

The City of Jouy Print

JUANACATLAN. Waterfalls
in Río Grande de Santiago,
W. central Mexico; 72 ft.
high

The Niagara Falls of Mex-
ico

JUBAYL, LEBANON. Small
village of Lebanon about
halfway between Tripolis
and Beirut

The City of Adonis

Justice, The Birthplace of
English. See: RUNNY-
MEDE, ENGLAND

-K-

KAFA (KAFFA). Region and
former province of S.W.
Ethiopia

The Home of the Coffee
Plant

KAGERA. River in N.W. Tan-
zania, E. Africa, flowing
from boundary of Ruanda-
Urundi to Lake Victoria;
about 430 mi. long

The Most Remote Headstream
of the Nile

Kamchatka, The Sea of. See:
BERING SEA

Kanarese, The Country of the.
See: CARNATIC, INDIA

KARA KUM. Desert area,
110,000 sq. mi., S. of
Lake Aral, stretching from
Caspian Sea to Amu Darya,
Soviet central Asia

The Black Desert

KARNAK, UNITED ARAB RE-
PUBLIC. Village of United
Arab Republic sit. on E.
bank of Nile R. about 300
mi. S. E. of Cairo

The Place of the Great
Temple

KASHMIR (CASHMERE), INDIA.
Valley sit. in central part
of Kashmir State, N. India

The Happy Valley
The Vale of Kashmir

KAUAI. Island of Hawaiian
group W. N. W. of Oahu

The Garden Island

KAVIR. See: DASHT-I-
KAVIR

KAZAN, U. S. S. R. Capital
of Tartar Soviet Socialist
Republic sit. on Kazanka
R. near its confluence with
the Volga

The Gate of Asia

KEA. See: MAUNA KEA

Keg of Europe. See: Powder
Keg of Europe, The

KELLS, IRELAND. Town of
W. Meath county, E. Ire-
land, 25 mi. W. of Drog-
heda

The Home of the Book of
Kells

KENT, ENGLAND. County of
S. E. England, 1525 sq. mi.

The Garden of England

Kentishmen, The Town of the.
See: CANTERBURY, ENG-
LAND

KERBELA, IRAQ. City of
central Iraq, 55 mi. S. S. W.
of Baghdad, sit. on edge of
desert W. of Hindiya R.

The City of Husain

KERCH, U. S. S. R. City, seaport and railway terminus
sit. in Crimean region of
U. S. S. R., 110 mi. N. E.
of Yalta

The Granary of Athens
Little Constantinople

Key of Christendom, The.
See: BUDA, HUNGARY

Key of India, The. See:
HERAT, AFGHANISTAN

Key of Russia, The. See:
SMOLENSK, U. S. S. R.

Key of the Dutch Seas, The.
See: FLUSHING, THE
NETHERLANDS

Key of the Gulf, The. See:
CUBA

Key of the Mediterranean,
The. See: GIBRALTAR

Key to the Highlands, The.

See: STIRLING, SCOTLAND

Khan, The City of the Great.
See: KHANBALIK, CHINA;
PEKING, CHINA

KHANBALIK, CHINA. Ancient
capital of China on site corresponding to modern Peking

The City of the Great Khan
The Great Capital

KHAZNA. See: EL KHAZNA

KHYBER PASS. Pass, about
33 mi. long, in Safed Koh
range on border between
Afghanistan and India, 10
mi. W. of Peshawar

The Gateway to Central
Asia
The Gateway to India

KIANG. See: SI-KIANG

KIANG RIVER. See: YANG-
TZE KIANG RIVER

KIDDERMINSTER, ENGLAND.
Municipal borough of Wor-
cestershire, W. central
England, 18 mi. W.S.W.
of Birmingham

The Home of Kiddermin-
ster Carpets

KIDRON WADI. Deep ravine
lying between Jerusalem
and the Mount of Olives
in Jordan-occupied Pales-
tine; site of Jewish and
Moslem cemetery

The Valley of Jehoshaphat

KIEV, U.S.S.R. City of
Ukrainian S.S.R. sit. on
right bank of Dnieper R.,
470 mi. S.W. of Mos-
cow

The Holy City
The Mother of Cities

KILIMA, MOUNT. Mountain
of E. Africa, 18,200 ft.
high

The Olympus of Ethiopia

KIMBERLEY, SOUTH AFRICA.
Town of N. Cape province,
Union of S. Africa, 86 mi.
N.W. of Bloemfontein

The Town of Diamonds
The World's Diamond Cen-
ter

KINCARDINE, SCOTLAND.
Maritime county of Scot-
land, sit. on North Sea

The Mearns

King, The City of the Great.
See: JERUSALEM, PAL-
ESTINE

King of Waters, The. See:
AMAZON RIVER; MISSIS-
SIPPI RIVER

Kingdom, The Flowery. See:
CHINA

Kingdom, The Lost. See:
INDIA

Kingdom, The Middle. See:
CHINA

Kingdom, The Middle Flowery.
See: CHINA

Kingdom of Burgundy, The.
See: ARLES, FRANCE

Kingdom of Mountains, The.
See: LAOS

Kingdom River, The Yang.
See: YANGTZE KIANG
RIVER

Kings. See: Valley of the Kings, The

Kings, The City of the. See: CASHEL, IRELAND; LIMA, PERU

Kings, The City of the Three. See: COLOGNE, GERMANY

Kings, The Inn of the. See: PARIS, FRANCE

Kings and Prophets, The Land of. See: PALESTINE

Kings Are Crowned, The City Where. See: TRONDHEIM, NORWAY

King's Crag, The. See: FIFE, SCOTLAND

KINSHA. Upper course of Yangtze R., China, down to Ipin, the junction point with the Min R.

The Golden Sand River

Kirk. See also: CHURCH

Kirk of Skulls, The. See: GOWRIE CHURCH

Kitchen. See: Hell's Kitchen

Klondike, The Arabian. See: LIBYA (LIBIA)

Knot, The City of the Gordian. See: GORDIUM, ASIA MINOR

Knuckle-Dusting City, The. See: LIVERPOOL, ENGLAND

KOH. See: SAFED KOH

KOMODO. Island E. of Soembawa Island and W. of Flores Island in Lesser Sunda Islands, Netherland Indies

The Home of the Dragon Lizard

KOREA. Country of E. Asia forming peninsula on Asiatic mainland; separated from Japan by Korea Strait

The Hermit Nation
The Land of the Morning Calm

KRALICKA, CZECHOSLOVAKIA. Town of N.E. Moravia, Czechoslovakia, E. of Prostejov

The City of the Brothers' Bible
The City of the Kralitz Bible

Kralitz Bible, The City of the. See: KRALICKA, CZECHOSLOVAKIA

Kru Coast, The. See: LIBERIA

KUM. See: KARA KUM

KURIL (KURILE) ISLANDS. Chain of 47 volcanic islands stretching from Hokkaido to Kamchatka Peninsula in Pacific Ocean

The Smoking Islands

KYOTO, JAPAN. City of Japan sit. in W. central Honshu 26 mi. N.N.E. of Osaka

The Paris of Japan

-L-

La Española (Little Spain). See: HISPANIOLA

LA GRANDE. See: CHAR-
TREUSE, LA GRANDE

La Manche (The Sleeve). See:
ENGLISH CHANNEL

LA PAZ, BOLIVIA. City of
La Paz dept., W. Bolivia,
sit. E. of Lake Titicaca;
elevation, 11,910 feet

The Peace of Ayacucho
The Sky High Capital

LA SALLE STREET, CHICAGO,
ILL. Street in financial district of Chicago

The Wall Street of Chicago

LACCADIVE ISLANDS. Group
of islands and coral reefs
in Arabian Sea, 200 mi.
off S.W. coast of India

The Hundred Thousand Is-
lands

Lace, The City of. See:
HONITON, ENGLAND

Ladder, The Giant's. See:
MOFFAT TUNNEL

Ladies' Mile, The. Drive in
Hyde Park, London, England

Ladies' Rock, The. See:
STIRLING, SCOTLAND

Lady Godiva, The Home of.
See: COVENTRY, ENGLAND

Lagoons, The City of the.
See: VENICE, ITALY

LAKE. See: AVERNUS,
LAKE; DRUMMOND, LAKE;
GEORGE, LAKE; GREEN-
WOOD LAKE; ITASKA,
LAKE; LUCERNE, LAKE
OF; MARACAIBO, LAKE;
SAIMAA, LAKE

Lake, The Great. See: TAI-
HU; TONLE SAP

Lake, The Western. See: SI-
HU

Lake District, The. Moun-
tainous region of N.W. England containing many lakes
and peaks; favorite resort
of English poets, including
Coleridge, Southey and Words-
worth

LAKE ERIE. See: Great
Lakes, The

Lake George, The Miniature.
See: GREENWOOD LAKE

Lake Horicon. See: GEORGE,
LAKE

LAKE HURON. See: Great
Lakes, The

LAKE MICHIGAN. See: Great
Lakes, The

Lake of Gennesaret, The. See:
GALILEE, SEA OF

Lake of Lot, The. See: DEAD
SEA

Lake of Oil, The. See: MAR-
ACAIBO, LAKE

Lake of the Dismal Swamp,
The. See: DRUMMOND,
LAKE

Lake of the Four Cantons, The.
See: LUCERNE, LAKE OF

Lake of the Four Forest Cantons, The. See: LUCERNE,
LAKE OF

Lake of the Thousand Lakes,
The. See: SAIMAA, LAKE

LAKE ONTARIO. See: Great

Lakes, The

LAKE SUPERIOR. See:
Great Lakes, The

Lakes. See: Great Lakes, The

Lakes, The Lake of the Thousand. See: LAIMAA,
LAKE

Lakes, The Land of a Thousand. See: FINLAND

Lakes, The Queen of the.
See: WINDERMERE

LAND. See: COCAGNE
(COCAIGNE), THE LAND
OF

Land. See: Debatable Land,
The; No Man's Land

Land, The Flowery. See:
CHINA

Land, The Forbidden. See:
TIBET

Land, Gimme Gimme. See:
LIBYA (LIBIA)

Land, The Great Thirst. See:
SOUTH AFRICA

Land, The Holy. See: ARABIA;
ELIS; INDIA; PALESTINE

Land, Potato. See: IRELAND
(EIRE)

Land, The Promised. See:
CANAAN, PALESTINE

Land, The Southern. See:
VIET NAM

Land, Wine. See: OENOTRIA,
ITALY

Land of a Thousand Lakes,
The. See: FINLAND

Land of Bondage, The. See:
EGYPT

Land of Cakes, The. See:
SCOTLAND

Land of Destiny, The. See:
IRELAND (EIRE)

Land of Frozen Time, The.
See: ANTARCTICA

Land of Giants, The. See:
PACIFIC NORTHWEST

Land of Green Ginger, The.
See: HULL, ENGLAND

Land of Health, The. See:
BADAJOZ, SPAIN

Land of Kings and Prophets,
The. See: PALESTINE

Land of Many Dragons, The.
See: VIET NAM

Land of Myrrh, The. See:
SABA

Land of Nod, The. Unknown
land on the east of Eden,
to which Cain fled (Genesis
IV)

Land of Prester John, The.
See: ETHIOPIA (ABYS-
SINIA)

Land of Promise, The. See:
CANAAN, PALESTINE

Land of Smile, The. See:
THAILAND

Land of Song, The. See:
ITALY

Land of the Boar, The. See:
GERMANY

Land of the Chaldees, The.
See: BABYLONIA

Land of the Fetish, The.
See: AFRICA

Land of the Free and the
Home of the Brave, The.
See: UNITED STATES OF
AMERICA

Land of the Hyacinth and Ruby,
The. See: SRI LANKA
(CEYLON)

Land of the Leal, The. See:
SCOTLAND

Land of the Leek, The. See:
WALES

Land of the Midnight Sun, The.
See: SCANDINAVIA

Land of the Morning Calm,
The. See: KOREA

Land of the Pharaohs, The.
See: EGYPT

Land of the Potato, The.
See: IRELAND (EIRE)

Land of the Pure, The. See:
PAKISTAN

Land of the Quetzal, The. See:
GUATEMALA

Land of the Rising Sun, The.
See: JAPAN

Land of the Rising Yen, The.
See: JAPAN

Land of the Rose, The. See:
ENGLAND

Land of the Shamrock, The.
See: IRELAND (EIRE)

Land of the Stars and Stripes,
The. See: UNITED STATES
OF AMERICA

Land of the Sun, The. See:
HINDUSTAN

Land of the Thistle, The. See:
SCOTLAND

Land of the Tulip, The. See:
NETHERLANDS, THE (HOL-
LAND)

Land of the Veda, The. See:
INDIA

Land of the White Elephant,
The. See: THAILAND

Land of the Free, The Flourishing. See: THAILAND

Land of Wisdom, The. See:
NORMANDY

Landfall, Columbus's First.
See: SAN SALVADOR (WAT-
LINGS ISLAND)

Land's End. See: CORNWALL,
ENGLAND; FINISTERRE,
CAPE

LAOS. Kingdom of Indochina,
sit. in W. and N.W. part,
W. of Vietnam

The Center of the World
The Kingdom of Mountains

LASHIO, BURMA. Town of N.
Shan states, E. Burma, 130
mi. N.E. of Mandalay

The Starting Point of the
Burma Road

LASSA. See: LHASA (LAS-
SA), TIBET

Last Moorish Stronghold in
Spain, The. See: GRANA-
DA, SPAIN

Last Sigh of the Moor, The.
Rocky eminence near Gra-
nada, Spain

LATAKIA, SYRIA. City of

Latakia Territory, W. Sy-
ria

The Home of Tobacco

Latitudes.   See: Horse Latitudes

Lavinia.   See: ITALY

Lazarus, The Home of.   See:
BETHANY, PALESTINE

Le Cap.   See: CAP-HAITIEN,
HAITI

League of Rhine Cities, The.
Collective designation of
Union of Mainz, Worms,
Oppenheim and others sit.
near Rhine R.

Leal, The Land of the.   See:
SCOTLAND

Leaning Tower, The.   Cylin-
drical tower at Pisa, Italy,
begun in 1174; several de-
grees off perpendicular

Leaning Tower, The City of
the.   See: PISA, ITALY

Leap, Sappho's.   See: DOU-
CATO, CAPE

Learned, The.   See: PADUA,
ITALY

Leek, The Land of the.   See:
WALES

Legendary Mouth of Hell, The.
See: AVERNUS, LAKE

LEIPZIG, GERMANY.   City of
the district of the same
name, E. Germany, 111
mi. S.W. of Berlin

Little Paris

LEITMERITZ, BOHEMIA.

Town of N. Bohemian pro-
vince, W. Czechoslovakia,
sit. on Elbe R. 35 mi.
N.N.W. of Prague

The Bohemian Paradise
The Paradise of Bohemia

LEMBERG (LVOV), U.S.S.R.
Commercial city of W.
Ukraine, U.S.S.R., 155
mi. S.W. of Lutsk

The Lion City

LEMURIA.   Mythical ancient
island or continent said to
have sunk beneath the surface of the Pacific Ocean

The Sunken Continent of
the Pacific

LENINGRAD, U.S.S.R.   Capital of Leningrad region of
Russian empire sit. on delta of Neva R. 400 mi. N.W.
of Moscow

The City of Palaces
The City of Perspectives
The City of Snow
The Palmyra of the North
A Window into Europe

Leonine City, The.   Portion
of Rome W. of Tiber R.
and N. of Trastevere; in-
cludes the Vatican

Lepers, The Island of the.
See: MOLOKAI

Leper's Well, The Isle of the.
See: STRONSAY

LESBOS.   Island in E. Aegean
Sea between Smyrna and
the Dardanelles off N.W.
coast of Turkey in Asia

The Island of Sappho

Levant, The Flower of the.

See: ZANTE

Level, Bedford. See: Fen
Country, The

LHASA (LASSA), TIBET. Bud-
dhist sacred city, capital of
Tibet, about 250 mi. N.E.
of Darjeeling

The Forbidden City
God's House
The Rome of Buddhism

LIBERIA. W. African re-
public sit. on coast of Atlantic Ocean S.E. of Sierra
Leone

The Grain Coast
The Kru Coast

Liberty, The Birthplace of.
See: INDEPENDENCE
HALL

Liberty, The Cradle of Amer-
ican. See: FANEUIL
HALL; INDEPENDENCE
HALL

Liberty Island. See: BED-
LOE'S ISLAND

LIBYA (LIBIA). Former colony of Italian colonial empire in N. Africa, sit. on
Mediterranean Sea

The Arabian Klondike
The Camel with the Feet
of Gold
Gimme Gimme Land
The Oil Sheikdom

LIEGE, BELGIUM. City of
E. Belgium, sit. at con-
fluence of Ourthe and
Meuse R.'s

The Birmingham of Bel-
gium

Light, The City of. See:

PARIS, FRANCE

Light of Greece, The. See:
CORINTH, GREECE

Light of the South, The. See:
SINGAPORE, ASIA

Lighthouse of El Salvador, The.
See: IZALCO

Lilies. See: Tank of the
Golden Lilies, The

Lilies, The City of. See:
FLORENCE, ITALY

LILLE, FRANCE. City of
Nord dept., N. France,
130 mi. N.N.E. of Paris

The City of Deportations

LIMA, PERU. Capital and
largest city of Peru, sit.
about 8 mi. E. of Callao

The City of the Kings

Limbo. Region supposed to
lie on the edge of Hell

Limbo. See also: Hell

LIMBURG, BELGIUM. Region
of N.E. Belgium where lim-
burger cheese originally
produced

The Home of Cheese

LIMERICK, IRELAND. City
of S.W. Ireland sit. on
Shannon R.

The City of the Violated
Treaty
The Stone of the Broken
Treaty

LIMPOPU RIVER. River in
E. part of S. Africa flow-
ing from W. of Pretoria to
Indian Ocean through Dela-

gosa Bay

The River of Crocodiles

LINDISFARNE. Peninsula of
N. E. coast of Northumber-
land, N. England; becomes
island at high tide

The Holy Island

Line, The Dividing. See:
YALU RIVER

Line, The Finish. See: JOHN
O' GROAT'S HOUSE

Lion, Britannia's. See: GI-
BRALTAR

Lion City, The. See: LEM-
BERG (LVOV), U.S.S.R.;
SINGAPORE, ASIA

Lion of the Sea, The. See:
GOOD HOPE, CAPE OF

LIPARI ISLANDS. Group of
seven volcanic islands in
Mediterranean Sea lying
off coast of Sicily N.W.
of Messina

The Aeolian Islands

Liqueur, The Home of Chartreuse. See: CHARTREUSE,
LA GRANDE

Little America. Base in Bay
of Whales, Ross Ice Shelf,
Antarctica, established by
explorer Richard E. Byrd
in 1928-1930

Little Britain. See: BRIT-
TANY, FRANCE

Little Constantinople. See:
KERCH, U.S.S.R.

Little England. See: BAR-
BADOS

Little Island, The Tight. See:
ENGLAND; TAIWAN (FOR-
MOSA)

Little Isle, The Tight. See:
ENGLAND

Little Paris. See: BRUSSELS,
BELGIUM; CAP-HAITIEN,
HAITI; LEIPZIG, GERMANY;
MILAN, ITALY

Little Spain (La Española).
See: HISPANIOLA

Little Switzerland of America,
The. See: REST ISLAND,
MINN.

Little Tibet. See: BALTISTAN,
INDIA

Little Venice. See: AMIENS,
FRANCE; ARENDAL, NOR-
WAY

LIVERPOOL, ENGLAND. City
of Lancashire, N.W. Eng-
land, sit. on Mersey estuary

The Knuckle-Dusting City

Lizard, The Home of the Dragon. See: KOMODO

LLOYDMINSTER, CANADA.
Town of Canada sit. on Al-
berta-Saskatchewan border
82 mi. W.N.W. of North
Battleford

The City of the All-British
Colony

Lloyd's Pulpit. See: Hugh
Lloyd's Pulpit

LOA. See: MAUNA LOA

LOCRI, ITALY. Ancient city
of Magna-Graecia, on E.
coast of Italy

The City of the Locrian Code

The City of Zaleucus

Locrian Code, The City of the.
See: LOCRI, ITALY

LOD, PALESTINE. Ancient
city of Judaea in Plain of
Sharon, now occupied by
Lydda, W. Palestine, 9
mi. S.E. of Jaffa

The City of St. George

Lode. See: Mother Lode,
The

LOIRE. See: INDRE-ET-
LOIRE

LOMAI VITI. Group of scat-
tered islands on W. side of
Koro Sea between Vanua
Levu and Kandavu, Fiji Islands in S.W. Pacific Ocean

Middle Fiji

LONDON. See: TOWER OF
LONDON

LONDON, ENGLAND. Capital
of United Kingdom of Great
Britain and N. Ireland sit.
in S.E. England on both sides
of Thames R. about 40 mi. inland from North Sea

The Big Smoke
Brute's City
The City
The City of Masts
The City of Smoke
Cocagne
The Great Smoke
Lubberland
The Modern Babylon
The Smoke

London of Fashion and Pleasure, The. See: WEST
END

Long Branch of Honolulu, The.

See: WAIKIKI BEACH

Long Branch of Philadelphia,
The. See: MAY, CAPE

Long Mountain. See: MAUNA
LOA

Long Rampart, The. See:
GREAT WALL, THE

Long Walk, The. Straight
avenue, about 3 mi. long
in Windsor Park, London

LONGWY, FRANCE. Com-
mune, Meurthe-et-Moselle
dept., N.E. France, 60
mi. N. of Nancy

The Iron Gate of France

Loop, The. The Point. Main
business section of Chicago,
Illinois

Lord, The Seat of the Mukden
War. See: MUKDEN, MAN-
CHURIA

Loretto of Switzerland, The.
See: EINSIEDELN, SWITZ-
ERLAND

Lost Causes, The City of.
See: GENEVA, SWITZER-
LAND

Lost Island. See: CEPHAL-
ONIA

Lost Kingdom, The. See:
INDIA

Lot, The Lake of. See: DEAD
SEA

Lot, The Sea of. See: DEAD
SEA

Lourdes of Wales, The. See:
HOLYWELL, WALES

Lovely Isle, The. See: COR-
FU

Low Countries, The. Low region bordering on North Sea
comprising modern Belgium,
Luxembourg and the Netherlands

Lower Turn, The. See: VUEL-
TA ABAJO

Lowlands, The. Low, level
region in S., central and
E. Scotland

Loyal City, The Ever. See:
OXFORD, ENGLAND

Lubberland. See: COCAGNE
(COCAIGNE), THE LAND
OF; LONDON, ENGLAND

LÜBECK, GERMANY. City
of Germany 35 mi. N.E.
of Hamburg

The Carthage of the North
The Queen of the Hanse

LUCCA, ITALY. Commune of
central Italy, 35 mi. W. by
N. of Florence

The Industrious

LUCERNE, LAKE OF. Lake
in central Switzerland en-
closed by cantons of Lu-
cerne, Schwyz, Uri and Unterwalden

The Lake of the Four Can-
tons
The Lake of the Four For-
est Cantons

LUCERNE, SWITZERLAND.
Commune of central
Switzerland on W. shore
of Lake of Lucerne, 25 mi.
S.S.W. of Zurich

The Metropolis of Tourists

LUCKNOW, INDIA. City of
India on right bank of Gumti
R. 270 mi. E.S.E. of Del-
hi

The City of Roses

LUNGKI, CHINA. Town and
county seat of county of
same name, Fukien province,
30 mi. W.N.W. of Amoy

The Home of Chu Hsi

Luther Towns, The. Collective appellation of German
towns of Eisleben, Coburg,
Erfurt, Magdeburg and Wittenberg

Luther's Sanctuary. See:
JENA, GERMANY

LVOV. See: LEMBERG
(LVOV), U.S.S.R.

LYSANIAS. See: TETRARCHY
OF LYSANIAS

-M-

McNamee Mountains, The Graham. See: SIERRA MADRE

MADEIRA. Largest of Madeira
Island group sit. in E. Atlantic Ocean off coast of
Morocco N. of Canary Is-
lands

The Home of Madeira Wines
The Isle De Dabney
The Pearl of the Atlantic

Madeira Wines, The Home of.
See: MADEIRA

MADRE. See: SIERRA MADRE

Magic Valley, The. Delta region of Rio Grande Valley in
S. Texas

Magna Charta, The Birthplace
of the. See: RUNNYMEDE,
ENGLAND

Magnetic Pole, The Home of
the North. See: BOOTHIA
PENINSULA, CANADA

Magnificent Mile, The. Appellation of N. Michigan
Ave., Chicago, Illinois

MAHANADI RIVER. River in
Indian Republic flowing from
Madhya Pradesh to Bay of
Bengal; about 512 mi. long

The Great River

Maid, The City of the. See:
ORLEANS, FRANCE

Maiden Town, The. See:
EDINBURGH, SCOTLAND

Maidens, The Castle of. See:
EDINBURGH, SCOTLAND

Main, The Spanish. See:
CARIBBEAN SEA

Main Stem, The. See: BROAD-
WAY, NEW YORK CITY

Main Street of America, The.
Federal highway, U.S. 66,
running from Chicago, Ill.,
to Los Angeles, Calif.

Mainland, The. See: UNITED
STATES OF AMERICA

MAINZ, GERMANY. Largest
city in state (formerly
grand duchy) of Hesse, W.
Germany, 18 mi. N.W. of
Darmstadt

The Birthplace of Printing

MALACCA, STRAIT OF. Strait
between S. Malay Peninsula
and island of Sumatra; about

500 mi. long

The Straits

MALAY ARCHIPELAGO. Is-
land group off S.E. coast of
Asia sit. between Pacific
and Indian Oceans

The Country in Search of a
Future
The Empire of the Islands
The Gardens of the Sun
The Islands of India

MALAY PENINSULA (MALAY-
SIA). Projection of Asiatic
mainland bounded by Gulf of
Siam, S. China Sea, Singapore Strait, Malacca Strait
and Bay of Bengal

The Golden Chersonese

Malmsey, The Home of. See:
MONEMVASIA, GREECE

Malvasia, The Home of. See:
MONEMVASIA, GREECE

MAM TOR. Mountain in Der-
byshire, England

The Shivering Mountain

Man, The Home of the Rho-
desian. See: BROKEN
HILL, AFRICA

MAN, ISLE OF. Island in
Irish Sea off N.W. coast of
England

The Isle of the Tailless Cat

Man, The Sick. See: TURKEY

Man Mountain, Old. See:
MOUNT HERMON

Man of Europe, The Sick. See:
TURKEY

Man of the East, The Sick.

See: TURKEY

Man of the Mountain, The Old.
See: CANNON MOUNTAIN

Man River, Old. See: MIS-
SISSIPPI RIVER

Manche, La (The Sleeve). See:
ENGLISH CHANNEL

Manchester, The Saxon. See:
CHEMNITZ, GERMANY

Manchester, The Soviet. See:
IVANOVO, RUSSIA

Manchester of Belgium, The.
See: GHENT, BELGIUM

Manchester of France, The.
See: ROUEN, FRANCE

Manchester of Prussia, The.
See: ELBERFELD, GER-
MANY

MANDALAY, BURMA. City of
central Upper Burma, 650
mi. up Irrawaddy R.

The Vatican of Buddhism

MANDEB. See: BAB EL
MANDEB

MANILA, P.I. City of S.W.
Luzon, Philippine Islands,
sit. on E. shore of Manila
Bay

The Pearl of the Orient

Man's Corner. See: Dead
Man's Corner

Man's Land. See: No Man's
Land

Many Dragons, The Land of.
See: VIET NAM

Many Marriages, The Town

of. See: GRETNA GREEN,
SCOTLAND

Mar del Sur. See: PACIFIC
OCEAN

MARACAIBO, LAKE. S. ex-
tension of Gulf of Venezuela
sit. in N.W. Venezuela; site
of many producing oil wells

The Lake of Oil

Mare Nostrum (Our Sea). See:
MEDITERRANEAN SEA

MARIA, SANTA. See: PUER-
TO DE SANTA MARIA, SPAIN

MARIANA ISLANDS. Group of
15 islands in Pacific Ocean
N. of Caroline Islands

The Stationary Island Aircraft Carriers

Marianne. See: FRANCE

MARINO. See: SAN MARINO

Mark. See: Spanish Mark,
The

MARKET STREET, SAN FRAN-
CISCO, CALIF. Main tho-
roughfare of San Francisco

The Path of Gold

Marriages, The Town of Many.
See: GRETNA GREEN,
SCOTLAND

Marsh, The Village of the.
See: BRUSSELS, BELGIUM

MARTIGUES, FRANCE. Com-
mune, Bouches-du-Rhône
dept., S.E. France, sit. on
Mediterranean Sea, 18 mi.
N.W. of Marseille

The Provençal Venice

MARTINIQUE. Island of the
Windward Islands, E. West
Indies

Empress Josephine's Island

Martyrdom, The Place of.
See: MESHED, IRAN

Martyrs. See: Pavement of
Martyrs, The

Martyrs, The Town of Twelve.
See: SCILLIUM, AFRICA

MAS A TIERRA. Island of
Juan Fernández group, off
coast of Chile, where Alexander Selkirk marooned

Crusoe's Island
Selkirk's Island

MASJID JABUD. Ancient ruin
sit. in city of Tabriz, capital of province of Azerbai-
jan, Iran

The Blue Mosque

MASJID JAMI. Mosque sit. in
Yezd, city and capital of
province of same name, Iran

The Friday Mosque

Massacre, The City of the.
See: CAWNPORE, INDIA

Masts, The City of. See:
LONDON, ENGLAND

Matches, The City of. See:
JÖNKÖPING, SWEDEN

MATTERHORN, THE. Moun-
tain peak in Pennine Alps
on Swiss-Italian border;
14,780 ft. high

Mont Cervin
Monte Silvio
The Tiger of the Alps

MAUNA KEA. Extinct volcano
on Hawaii Island, state of
Hawaii; 13,823 ft. high

White Mountain

MAUNA LOA. Volcano on
Hawaii Island, state of Hawaii; 13,684 ft. high

Long Mountain

MAVRO NERO. River of N.E.
Arcadia, Greece, regarded
by ancient Greeks as the
entrance to the lower world

The River Styx

MAY, CAPE. Cape of S. New
Jersey separating Delaware
Bay from Atlantic Ocean

The Long Branch of Phila-
delphia

Meadow, The Great. See:
GRAND PRE, NOVA SCOTIA

Mearns, The. See: KINCAR-
DINE, SCOTLAND

MECCA, SAUDI ARABIA. One
of two capitals of Saudi Arabia located about 45 mi. E.
of Jidda; birthplace of Mohammed

The Holy City of Islam

Mecca of Spain, The. See:
SANTIAGO DE COMPOSTELA

Medea's Home. See: COL-
CHIS

MEDINA, SAUDI ARABIA. City
of Saudi Arabia sit. about
820 mi. S.E. of Damascus;
site of tomb of Mohammed

The City of the Apostle of
God

The City of the Prophet
The City of the Refuge
The Holy City of Islam
The Prophet's City

Mediterranean, The Gate of
the. See: GIBRALTAR,
STRAIT OF

Mediterranean, The Key of
the. See: GIBRALTAR

Mediterranean, The Pearl of
the. See: ALEXANDRIA,
EGYPT

Mediterranean of Brazil, The.
See: AMAZON RIVER

Mediterranean of the North,
The. See: BALTIC SEA

MEDITERRANEAN SEA. In-
land sea between continents
of Europe, Asia and Africa

The Cradle of Civilization
The Great Sea
Mare Nostrum (Our Sea)

Meet, Where Past and Present.
See: CAMBODIA

Meeting, A Place of. See:
TORONTO, CANADA

MELOS, GREECE. Ruined
city on Island of Melos,
sit. in Cyclades dept.,
Greece, where statue
"Venus de Milo" was discov-
ered in 1820

The City of Venus

Men, The Home of the Wise.
See: GOTHAM, ENGLAND

Men, The Republic of Free.
See: TAIWAN (FORMOSA)

Mercy, The House of. See:
BETHESDA

MEROË. Now ruined ancient
city sit. on E. bank of
Nile R.; capital of Ethi-
opian kings, of Nubia and
of the Meroitic kingdom
(c. 700 B.C. to c. 350
A.D.)

The Queen of the World

Merrie England. See: ENG-
LAND

Merry England. See: ENG-
LAND

MESHED, IRAN. City of N.E.
Iran sit. on tributary of
Hari-Rud R., 460 mi. N.E.
of Teheran

The Place of Martyrdom

MESOPOTAMIA. Region in
S.W. Asia between Euphrates
and Tigris R.'s, extending
from Persian Gulf to S.
Armenia (now part of Tur-
key)

The Country Between Two
Rivers

Metropolis of Flora, The.
See: ARANJUEZ, SPAIN

Metropolis of the Orient, The
Industrial. See: OSAKA,
JAPAN

Metropolis of Tourists, The.
See: LUCERNE, SWITZ-
ERLAND

MEXICO. Federated republic
of N. America S. of United
States of America

Montezuma's Realm
South of the Border

Mexico, The Niagara Falls of.
See: JUANACATLAN

MEXICO CITY, MEXICO. Capital of Republic of Mexico, 264 mi. W.N.W. of Veracruz

Cactus on a Stone

Miami Beach East. See: TEL AVIV, ISRAEL

MICHIGAN, LAKE. See: Great Lakes, The

Middle Fiji. See: LOMAI VITI

Middle Flowery Kingdom, The. See: CHINA

Middle Kingdom, The. See: CHINA

Midnight Sun, The Land of the. See: SCANDINAVIA

Might, A Symbol of Modern. See: GIBRALTAR

MILAN, ITALY. Capital and city of Milano province, Lombardy, Italy, 300 mi. N.W. of Rome

Little Paris

Mile. See: Ladies' Mile, The; Magnificent Mile, The

Mile of Christmas Trees, The. Appellation of Santa Rosa Avenue, Pasadena, Calif.

Million Dreams, The City of a. See: VIENNA, AUSTRIA

Millionaire's Resort, The. See: JEKYLL ISLAND

Millionaire's Row. Appellation of Orange Drive and Oak Knoll, Pasadena, Calif. Also highway on E. bank of Hudson R., New York

City to Yonkers, N.Y.

Mine of Europe, The Gold. See: TRANSYLVANIA

Mines, The City of the. See: IGLESIAS, SARDINIA

Miniature Lake George, The. See: GREENWOOD LAKE

Missions, The Mother of Christian. See: ANTIOCH, ASIA MINOR

MISSISSIPPI RIVER. Navigable R. in central United States, 2470 mi. to head of the Passes; 3988 mi. long if measured from headwaters of Missouri R.

The Backbone of the Confederacy
The Father of Waters
The Great River
The Hudson River of the West
The King of Waters
Old Man River

Mississippi River, The Infant. See: ITASCA, LAKE

Mississippi River, Nicollet's Infant. See: ITASCA, LAKE

Mississippi, The Source of the. See: ITASCA, LAKE

MISSOURI RIVER. Principal branch of Mississippi R.; formed by confluence of Jefferson, Gallatin and Madison R.'s in Montana; about 2475 mi. long

Big Muddy River

Mistress of the Adriatic, The. See: VENICE, ITALY

Mistress of the Seas, The.

See: GREAT BRITAIN

Mistress of the World, The.
See: ROME, ITALY

Model Province, The. See:
SHANSI, CHINA

Modern Athens, The. See:
EDINBURGH, SCOTLAND

Modern Babylon, The. See:
LONDON, ENGLAND

Modern Might, A Symbol of.
See: GIBRALTAR

Modern Paternò, The. See:
HYBLA, SICILY

MOFFAT TUNNEL. Railroad
tunnel 6.1 mi. long through
James Peak, N. central
Colorado, about 50 mi. N.W.
of Denver

The Giant's Ladder

Mogen, Hogen. See: NETHER-
LANDS, THE (HOLLAND)

Mohammedan Athens, The.
See: BAGHDAD, IRAQ

MOLOKAI. One of Hawaiian
Islands sit. in Pacific Ocean
about 10 mi. N.W. of Maui;
site of leper colony on N.
coast

The Island of the Lepers

MOLUCCAS. Group of islands
in innermost part of Malay
Archipelago forming part of
Indonesia

The Spice Islands

Monarchy, The Dual. See:
AUSTRIA-HUNGARY

MONEMVASIA, GREECE.

Village on small island off
coast of S.E. Laconia dept.,
S.E. Peloponnesus, Greece

The Home of Malmsey
The Home of Malvasia

Money, The Island of Stone.
See: YAP

Money Pit, The. Site of pos-
sible buried treasure on
Oak Island in Atlantic Ocean,
off coast of Nova Scotia

Mongolia, The Gate to. See:
WANCHUAN, MONGOLIA

Mont. See also: MONTE;
Monte; MOUNT; Mount;
MOUNTAIN; Mountain

Mont Cervin. See: MATTER-
HORN, THE

Montcalm's Waterloo. See:
PLAINS OF ABRAHAM

MONTE. See: CORNO, MONTE

MONTE (Monte). See also:
Mont; MOUNT; Mount;
MOUNTAIN; Mountain

Monte Silvio. See: MATTER-
HORN, THE

MONTENEGRO. Former king-
dom, now republic of the
Federated Republic of Yugoslavia sit. in S.W. portion
of that country

The Black Mountain

Montenegro of Africa, The.
See: ETHIOPIA (ABYS-
SINIA)

Montenegro of Sumatra, The.
See: ACHIN, N. SUMATRA

Montezuma's Realm. See:
MEXICO

MONTGOMERY STREET, SAN
FRANCISCO, CALIF.   Street
in financial district of San
Francisco

The Wall Street of the West

Monticello.   Estate and residence of Thomas Jefferson,
sit. 3 mi. S.E. of Char-
lottesville, Va.

Montpellier of Australia, The.
See: BRISBANE, AUSTRALIA

MONTREAL, CANADA.   City of
Hochelaga County, sit. on
S.E. Montreal Island, S. Que-
bec, Canada, on N. bank of
St. Lawrence R.

The Island City

MONTSERRAT.   Volcanic island
of British colony of Leeward
Islands, British W. Indies,
27 mi. S.W. of Antigua

The Emerald Island of the
West

Moon, The Mountains of the.
See: RUWENZORI, MOUNT

Moonrakers, The Home of the.
See: WILTSHIRE

Moonshot of 1883, The.   See:
BROOKLYN BRIDGE

Moor.   See: Last Sigh of the
Moor, The

Moorish Stronghold in Spain,
The Last.   See: GRANADA,
SPAIN

MORAYSHIRE.   County of N.E.
Scotland

The Garden of Scotland

Morning Calm, The Land of

the.   See: KOREA

MOROCCO.   African kingdom
on Atlantic Ocean and Mediterranean Sea

The Far West

Moroland.   Islands, sit. S. of
Philippine Islands, where
Moros live; primarily Min-
danao and the Sulu Archi-
pelago

MOSCOW, RUSSIA.   Capital,
inland port and largest city
of U.S.S.R., sit. on Mos-
kva R., 400 mi. S.E. of
Leningrad

The Heart of the Empire
The Holy City
The Holy Mother of the Rus-
sians
The Port of Five Seas
The White-Stoned

Moses, The Mountain of.   See:
JEBEL MUSA

Moses, Smoking.   See: SHI-
SHALDIN

Mosque, The Blue.   See: MAS-
JID JABUD

Mosque, The Friday.   See:
MASJID JAMI

Mosques, The City of.   See:
DAMASCUS, SYRIA

Most English Town in South
Africa, The.   See: GRA-
HAMSTOWN, SOUTH AFRICA

Most Remote Headstream of
the Nile, The.   See: KA-
GERA

Mother, The Goddess.   See:
EVEREST, MOUNT

Mother Lode, The.   Primary

belt of gold-bearing quartz in W. foothills of Sierra Nevada mountains, California

Mother of Books, The. See: ALEXANDRIA, EGYPT

Mother of Christian Missions, The. See: ANTIOCH, ASIA MINOR

Mother of Cities, The. See: BALKH, BACTRIA; KIEV, U.S.S.R.

Mother of Diets, The. See: WORMS, GERMANY

Mother of Parliaments, The. See: ENGLAND

Mother of the Potteries, The. See: BURSLEM, ENGLAND

Mother of the Russians, The Holy. See: MOSCOW, RUSSIA

Mother of the World, The Goddess. See: EVEREST, MOUNT

MOUNT. See: ATHOS, MOUNT; DAVIDSON, MOUNT; ETNA, MOUNT; EVEREST, MOUNT; FUJI (FUJIYAMA), MOUNT; HERMON, MOUNT; KILIMA, MOUNT; RUWENZORI, MOUNT; TITANO, MOUNT

MOUNT (Mount). See also: Mont; MONTE; Monte; MOUNTAIN; Mountain

Mount Fuji. See: FUJI (FUJIYAMA), MOUNT

Mount Sinai. See: JEBEL MUSA

Mount Vernon. Home and burial place of George Washington, sit. in Fairfax County, Virginia, on Potomac R., 15 mi. below Washington, D.C.

MOUNTAIN. See: CANNON MOUNTAIN

MOUNTAIN (Mountain). See also: Mont; MONTE; Monte; MOUNT; Mount; MOUNTAINS; Mountains

Mountain, The Black. See: MONTENEGRO

Mountain, China's Sacred. See: TAI-SHAN

Mountain, The Goddess. See: EVEREST, MOUNT

Mountain, The Holy. See: ATHOS, MOUNT

Mountain, Long. See: MAUNA LOA

Mountain, The Mysterious. See: EVEREST, MOUNT

Mountain, Old Man. See: HERMON, MOUNT

Mountain, The Old Man of the. See: CANNON MOUNTAIN

Mountain, The Profile. See: CANNON MOUNTAIN

Mountain, The Sacred. See: FUJI (FUJIYAMA), MOUNT

Mountain, The Shivering. See: MAM TOR

Mountain, The Smoking. See: POPOCATEPETL

Mountain, Snow. See: HERMON, MOUNT

Mountain, The Snow Covered.

See: ILLIMANI

Mountain, White. See: MAU-
NA KEA; SAFED KOH

Mountain of Fire, The. See:
ETNA, MOUNT

Mountain of Moses, The. See:
JEBEL MUSA

Mountain of Silver, The. See:
DAVIDSON, MOUNT

Mountain of Sorrow, The. See:
EDINBURGH, SCOTLAND

Mountain of Ten Thousand Ancients, The. See: WAN-
SHOU-SHAN

Mountain of Terrors, The.
See: SCHRECKHORN,
GROSS

Mountain of Three Peaks, The.
See: TITANO, MOUNT

Mountain Republic, The. Tem-
porary subdivision (1921-
27) of S.E. Soviet Russia,
Europe, later replaced by
certain autonomous repub-
lics on N. slopes of Cau-
casus Mountains

MOUNTAINS. See: ANDES
MOUNTAINS; BALKAN
MOUNTAINS; GREAT
SMOKY MOUNTAINS; HARZ
MOUNTAINS; IDA MOUN-
TAINS; ROCKY MOUNTAINS;
WHITE MOUNTAINS

MOUNTAINS (Mountains). See
also: MOUNTAIN; Mountain

Mountains, The Celestial. See:
TIEN-SHAN

Mountains, The Graham McNamee.
See: SIERRA MADRE

Mountains, The Helvetian.

See: ALPS, SWISS

Mountains, The Kingdom of.
See: LAOS

Mountains of the Moon, The.
See: RUWENZORI, MOUNT

Mourning. See: Field of
Mourning, The

Mouth of Hell, The Legendary.
See: AVERNUS, LAKE

Mouth of the Han River, The.
See: HANKOW, CHINA

Mouth of the Wells of Itzá,
The. See: CHICHEN ITZA

Movement, The Birthplace of
the Cooperative. See: ROCH-
DALE, ENGLAND

Muddy River, Big. See: MIS-
SOURI RIVER

Muddy Waters, The Place of
the. See: SHABARAKH
USU

MUKDEN, MANCHURIA. Cap-
ital of Manchuria sit. in
province of Liaotung, 425
mi. N.E. of Peking

The Seat of the Mukden
War Lord

MUNICH, BAVARIA. Capital
of Bavaria, W. Germany,
sit. on Isar R. 33 mi.
E.S.E. of Augsburg

The Town of the Beer Hall
Putsch

MUSA. See: JEBEL MUSA

Muses, The Abode of Apollo
and the. See: HELICON

Museum City, The. See:
NOVGOROD, RUSSIA

Mutineers, The Island of.
See: PITCAIRN ISLAND

MYCENAE, GREECE. Ancient ruined city of Argolis and Corinth dept., N.E. Peloponnesus, Greece, sit. about 7 mi. N. of Argos

Agamemnon's Capital

Myrrh, The Land of. See: SABA

Mysteries, The Seat of the Eleusinian. See: ELEUSIS, GREECE

Mysterious Mountain, The. See: EVEREST, MOUNT

-N-

NAGA HILLS. Hill region of Assam and Burma; part of N. Arakan Yoma System

The Home of the Head-Hunters

NAM. See: VIET NAM

Nameless City, The. See: ROME, ITALY

NANKING, CHINA. Independent municipality of Kiangsu province, China, sit. on Yangtze R., 170 mi. W.N.W. of Shanghai

The Southern Capital

Naples. See: Versailles of Naples, The

NAPLES, ITALY. Capital and seaport of Napoli province, Italy, 120 mi. S.E. of Rome

The Beautiful

The New City

NATCHEZ TRACE. Old road reaching from Nashville, Tenn. to Natchez, Miss.; over 500 mi. long

The Devil's Backbone

Nation, The Hermit. See: KOREA

Nation, The Island. See: SINGAPORE ISLAND

Nation at the Crossroads, The. See: BURMA

NATIONAL PARK. See: YELLOWSTONE NATIONAL PARK

Nations, The Niobe of. See: ROME, ITALY

Nations, The Playground of Two. See: GREAT SMOKY MOUNTAINS

Native States, The. See: INDIAN STATES

Nativity, The Site of the. See: BETHLEHEM, PALESTINE

Navigators Islands. See: SAMOA ISLANDS

Near East, The. Appellation of the Balkan states and the countries of S.W. Asia (the countries of the Arabian Peninsula)

Near East, The. See also: East, The; Far East, The

Neckar Wine, The Home of. See: ESSLINGEN, WEST GERMANY

Negro Population of the United States, The Capital of the. See: HARLEM

Nergal, The City of. See:
CUTHAH, BABYLONIA

NERO. See: MAVRO NERO

Nest. See: Hornet's Nest,
The

NETHERLANDS, THE (HOL-
LAND). Constitutional
monarchy of Europe sit.
in W. central region of
continental mainland on
North Sea

The Country of Paradoxes
Hogen-Mogen
The Land of the Tulip

Neutral Territories, The. Two
triangular-shaped areas of
desert land on N.E. boundary of Saudi Arabia, not in-
cluded in any sovereignty
because of incomplete boundary adjustments

New Albion. Portion of American Pacific Coast including
N. California, Oregon and
northward region

NEW CALEDONIA. Island
overseas territory of France
sit. in S.W. Pacific Ocean

New Cal

New City, The. See: NA-
PLES, ITALY

New England, The Top of.
See: WHITE MOUNTAINS

New Holland. See: AUS-
TRALIA

New Town, The. See:
CARTHAGE, NORTH AF-
RICA

New World, The. Land of
W. Hemisphere comprising

N. and S. America

New World, The Dardanelles
of the. See: DETROIT
RIVER

New World, The Gibraltar of
the. See: DIAMOND, CAPE

New World, The Sea of the.
See: CARIBBEAN SEA

New World of Tomorrow, The.
See: EQUATORIAL AFRICA

New York, The Hub of. See:
COLUMBUS CIRCLE, NEW
YORK CITY

NEWGATE PRISON. Prison
sit. in London, England

Whit's Palace

NEWGATE PRISON. See also:
Birdcage Walk

Newmarket of America, The.
Appellation of Monmouth
Park Race Track, New Jer-
sey

Niagara Falls of Mexico, The.
See: JUANACATLAN

Niagara of the East, The.
Waterfalls on St. John
R., N.E. Maine

Nice of the Adriatic, The. See:
ABBAZIA, CROATIA

Nicollet's Infant Mississippi
River. See: ITASCA, LAKE

Nights, The City of the Arabian.
See: BAGHDAD, IRAQ

Nile, The American. See:
SAINT JOHN'S RIVER

Nile, The Most Remote Headstream of the. See: KA-
GERA

NILE RIVER. Largest river in Africa; flowing from Kagera, its most remote headstream, to Mediterranean Sea; about 4000 mi. long

The Giant
The Great River

Niobe of Nations, The. See: ROME, ITALY

NIUAFOO. Island of N. Tonga Archipelago, central Pacific Ocean, 400 mi. N. of Tonga-tabu

Tin Can Island

No Man's Land. Area between Allied and German trenches in World War I

No Return, The Isle of. See: ALCATRAZ ISLAND

No Return, The River of. See: BIG SALMON RIVER

Nod. See: Land of Nod, The

NOOSA HEADS. String of surfing beaches on coast of Australia 60 mi. N. of Brisbane on Pacific Ocean

Australia's Sunshine Coast

NORMANDY. Region of N.W. France, sit. on English Channel

The Land of Wisdom

North, The Athens of the. See: COPENHAGEN, DENMARK; EDINBURGH, SCOTLAND

North, The Carthage of the. See: LÜBECK, GERMANY

North, The Farthest. See: HAMMERFEST, NORWAY

North, The Florence of the. See: DRESDEN, GERMANY

North, The Mediterranean of the. See: BALTIC SEA

North, The Palmyra of the. See: LENINGRAD, U.S.S.R.

North, The Queen of the. See: EDINBURGH, SCOTLAND

North, The Rome of the. See: COLOGNE, GERMANY

North America, The Backbone of. See: ROCKY MOUNTAINS

North America, The Gibraltar of. See: SAINT JOHN'S NEWFOUNDLAND

North Atlantic, The Fisherman's Paradise of the. See: BLOCK ISLAND

North Britain. See: SCOTLAND

North Magnetic Pole, The Home of the. See: BOOTHIA PENINSULA, CANADA

North of the Border. See: CANADA

NORTH POLE. N. extremity of earth's axis at 90° N. latitude

The Top of the World

Northern Athens, The. See: EDINBURGH, SCOTLAND

Northern Bear, The. See: RUSSIA

Northern Capital, The. See: PEKING, CHINA

Northern Giant, The. See:
RUSSIA

Northern Seas, The Circuit of
the. See: HOKKAIDO, JA-
PAN

NORTHWEST. See: PACIFIC
NORTHWEST

Norway, The East Valley of.
See: OSTERDAL

Norway's Farthest West. See:
STEINSÖY, NORWAY

NORWICH, ENGLAND. City
of E. England; capital of
Norfolk County, 114 mi.
N.N.E. of London

The Cathedral City

Nose Ridge, Bloody. See:
UMURBROGOL

Nostrum, Mare (Our Sea). See:
MEDITERRANEAN SEA

Notariate, The Fountain of
Doctors of the. See:
FLORENCE, ITALY

NOTTINGHAM, ENGLAND.
City of Nottinghamshire,
England, 125 mi. N.N.W.
of London

The Home of the Caves

Nova Scotia, The Inland Gateway to. See: AMHERST,
NOVA SCOTIA

NOVGOROD, RUSSIA. Rus-
sian city, capital of region
of same name, 110 mi.
S.S.E. of Leningrad

The Museum City

NOVOSIBIRSK, RUSSIA. City
of Novosibirsk region, sit.

in S. part of Soviet Russia,
Asia, about 390 mi. E. of
Omsk

The Chicago of Siberia

No. 10 Downing Street. Official residence of British
Prime Minister in W. end
of London, England, where
cabinet meetings often held

Numidian Pompeii, The. See:
TIMGAD, ALGERIA

-O-

Oak, The Place of the Royal.
See: BOSCOBEL, ENGLAND

OBERAMMERGAU, BAVARIA.
Village of Upper Bavaria,
W. Germany, 42 mi. S.S.W.
of Munich

The Home of the Passion
Play

OCEAN. See: ATLANTIC
OCEAN; PACIFIC OCEAN

Ocean, The Southern. See:
PACIFIC OCEAN

Ocean, The Western. See:
ATLANTIC OCEAN

Oceania. Collective name for
islands of central and S.
Pacific Ocean, including
Melanesia, Micronesia,
Polynesia and (sometimes)
Australia, the Malay Archi-
pelago and New Zealand

ODESSA, RUSSIA. Seaport
city of U.S.S.R. sit. on
Odessa Bay N.E. of mouth
of Dniester R.

The Queen of the Black Sea

O'Donnell's Country. See:
DONEGAL, IRELAND

Odysseus, The Home of. See:
ITHACA

OENOTRIA, ITALY. Ancient
region of S. Italy; comprised
modern Calabria and Luciana

Wine Land

Oil, The Lake of. See: MAR-
ACAIBO, LAKE

Oil Capital of the World, The.
See: DHAHRAN, SAUDI
ARABIA

Oil Sheikdom, The. See:
LIBYA (LIBIA)

OIMYAKON, RUSSIA. Town
of S. E. Yakutsk Republic,
Soviet Russia, Asia, on upper Indigirka R.

The Icebox of Siberia

Old. See also: Auld; Owld

Old Colony, The. See: PLY-
MOUTH COLONY

Old Country, The. See: AUS-
TRALIA; ENGLAND; GREAT
BRITAIN

Old Gib. See: GIBRALTAR

Old Man Mountain. See:
HERMON, MOUNT

Old Man of the Mountain, The.
See: CANNON MOUNTAIN

Old Man River. See: MIS-
SISSIPPI RIVER

OLD SOUTH CHURCH. Church
located at Washington and
Milk Streets, Boston, Mass.

Old South

The Sanctuary of Freedom

Old World, The. Appellation
of E. Hemisphere

Oldest Settlement, The. See:
SANTO DOMINGO, HISPAN-
IOLA

Oldest State in Europe, The.
See: SAN MARINO

Olympic Games, The Country
of the. See: ELIS

Olympus of Ethiopia, The.
See: KILIMA, MOUNT

Oman's Sea. See: PERSIAN
GULF

Once-Royal City of Herod, The.
See: TIBERIAS, PALESTINE

One of the Seven Wonders of
Wales. See: SAINT GILES'S
CHURCH

ONTARIO, LAKE. See: Great
Lakes, The

OPORTO, PORTUGAL. Sea-
port city of Pórto district,
N. W. Portugal, 170 mi.
N. E. of Lisbon

The Home of Port Wine

Oracle, The Home of the.
See: DELPHI, GREECE

Orchard of Denmark, The.
See: FALSTER ISLAND

Orchard of Ireland, The. See:
ARMAGH, IRELAND

Orient, The. See: Far East,
The

Orient, The Citadel of the. See:
SINGAPORE ISLAND

Orient, The Hub of the. See:

TAIPEI, FORMOSA

Orient, The Industrial Metropolis of the. See: OSAKA, JAPAN

Orient, The Pearl of the. See: MANILA, P. I.

ORLEANS, FRANCE. Capital city of department of Loriet, France, 77 mi. S. W. of Paris

The City of the Maid

ORPINGTON, ENGLAND. Town of Kent, S. E. England, 14 mi. S. E. of London

The Town of the Orpington Fowl

OSAKA, JAPAN. Seaport city of Honshu, Japan, sit. on N. E. shore of Osaka Bay

The Industrial Metropolis of the Orient
The Paris of Japan
The Venice of Japan

Osiris, The City of. See: BUSIRIS, EGYPT

OSTERDAL. Valley of Norway parallel to Swedish border

The East Valley of Norway

Other Islands, The. See: OVERIGE EILANDEN

OUDH, INDIA. N. E. portion of United Provinces of Agra and Oudh in N. India; former province of British India

The Garden of India

Our Sea (Mare Nostrum). See: MEDITERRANEAN SEA

OVERIGE EILANDEN. Former division of the Outer Provinces of Netherlands Indies, including Bali and Lombok, Timor and Moluccas residencies

The Other Islands

Owld. See also: Auld; Old

Owld Country, The. See: IRELAND (EIRE)

Owld Sod, The. See: IRELAND (EIRE)

Ox, The Ford of the. See: BOSPORUS

Ox-Bow Route. See: Great Ox-Bow Route, The

OXFORD, ENGLAND. City of Oxfordshire, E. England; site of Oxford University

The Ever-Loyal City

Oyster Island. See: ELLIS ISLAND

-P-

Pacific, The Gettysburg of the. See: TRUK

Pacific, The Gibraltar of the. See: PEARL HARBOR

Pacific, The Grand Canyon of the. See: WAIMEA CANYON

Pacific, The Paradise of the. See: HAWAIIAN ISLANDS

Pacific, The Sunken Continent of the. See: LEMURIA

PACIFIC NORTHWEST. Area of continental United States

including N. California,
Oregon, Washington and
parts of Idaho

The Land of Giants

PACIFIC OCEAN. Body of
water lying between Amer-
ica, Australia, Malay archipelago and Asia mainland;
largest and deepest of the
five oceans

Mar del Sur
The Peaceful
The South Sea
The Southern Ocean

PADUA, ITALY. City and
capital of Padua province,
Italy, sit. 22 mi. W. of
Venice

The Learned

Paganism, The Stronghold of.
See: HARZ MOUNTAINS

Pain, The Way of. See: Via
Dolorosa (The Way of Pain)

PAKISTAN. Confederation of
regions formerly in Indian
Empire, established in 1947
as self-governing dominion
of British Commonwealth of
Nations

The Land of the Pure

Palace, Whit's. See: NEW-
GATE PRISON

Palace, The Yankee. See:
WHITE HOUSE

Palaces, The City of. See:
CALCUTTA, INDIA; EDIN-
BURGH, SCOTLAND; GENOA,
ITALY; LENINGRAD, U.S.S.R.;
PARIS, FRANCE; ROME,
ITALY

Pale. See: English Pale, The

PALESTINE. Ancient region
of W. Asia bordering E.
Mediterranean Sea

The Holy Land
The Land of Kings and Prophets

Palmyra of the Deccan, The.
See: BIJAPUR, INDIA

Palmyra of the North, The.
See: LENINGRAD, U.S.S.R.

PAMIR. High altitude region
of central Asia, mostly sit.
in Tadzhik, S.S.R., Soviet
central Asia

The Roof of the World

Pan Hill, Salt. See: ZOUT-
PANSBERG

PANAMA. Republic of central
South America occupying
Isthmus of Panama

The Switzerland of Central
. America
The Switzerland of This
Hemisphere

PANAMA CANAL. Ship canal
extending S.E. across Isthmus of Panama from Colón
on Caribbean Sea to Balboa
on Bay of Panama

The Big Ditch
The Busy Ditch

Panhandle, The. Areas in
N.W. Oklahoma, N.W. Tex-
as, N. West Virginia, N.
Idaho and elsewhere in United
States

Pantheon of the British, The.
See: COLLEGIATE CHURCH
OF ST. PETER

Papal States, The. States of the
Church, The. Temporal domain

of the Pope in central Italy
from 755 to 1870

Paradise. See: HAWAIIAN
ISLANDS

Paradise, The Bohemian. See:
LEITMERITZ, BOHEMIA

Paradise, The Dutch. See:
GELDERLAND, E. NETHER-
LANDS

Paradise, The Golfer's. See:
SAINT ANDREWS

Paradise, The Portuguese.
See: SINTRA, PORTUGAL

Paradise of Bohemia, The.
See: LEITMERITZ, BO-
HEMIA

Paradise of the North Atlantic,
The Fisherman's. See:
BLOCK ISLAND

Paradise of the Pacific, The.
See: HAWAIIAN ISLANDS

Paradise of the World, The.
See: CONGO, REPUBLIC
OF THE

Paradoxes, The Country of.
See: NETHERLANDS, THE
(HOLLAND)

PARAGUAY. Republic of central S. America N. E. of
Argentina

The Heroin Crossroads of
South America

PARIS, FRANCE. Capital city
of France sit. on both banks
of Seine R. about 110 mi.
E.S.E. of Le Havre

The City of Light
The City of Palaces
The Inn of the Kings

Paris, Little. See: BRUS-
SELS, BELGIUM; CAP-
HAITIEN, HAITI; LEIPZIG,
GERMANY; MILAN, ITALY

Paris of Eastern Europe, The.
See: VIENNA, AUSTRIA

Paris of Japan, The. See:
KYOTO, JAPAN; OSAKA,
JAPAN

Paris of the Ancient World,
The. See: CORINTH,
GREECE

PARK. See: CLARENDON
PARK, ENGLAND; YELLOW-
STONE NATIONAL PARK

Parliaments, The Mother of.
See: ENGLAND

PARMA, ITALY. Capital city
of Parma province, Italy,
sit. 55 mi. W.N.W. of
Bologna

The Home of Parmesan
Cheese

Parmesan Cheese, The Home
of. See: PARMA, ITALY

Parnassus of Japan, The. See:
FUJI (FUJIYAMA), MOUNT

Part of a Province, The Fourth.
See: TETRARCHY OF LY-
SANIAS

PASS. See: BRENNER PASS;
KHYBER PASS; PAZ

Passion Play, The Home of
the. See: OBERAMMER-
GAU, BAVARIA

Past and Present Meet, Where.
See: CAMBODIA

Paternò, The Modern. See:
HYBLA, SICILY

Path of Gold, The. See: MAR-
KET STREET, SAN FRAN-
CISCO, CALIF.

Pavement of Martyrs, The.
Place on battlefield near
Tours, France, where
Charles Martel defeated
Saracens in the year 732

PAVIA, ITALY. Commune of
Pavia province, Lombardy,
N. Italy, sit. on Ticino R.
19 mi. S. of Milan

The City of a Hundred
Towers

PAZ. See: LA PAZ, BOLIVIA;
PASS

Peace, The City of. See:
BAGHDAD, IRAQ; GENEVA,
SWITZERLAND; JERUSA-
LEM, PALESTINE

Peace, The Isle of. See:
AQUIDNECK ISLAND

Peace of Ayacucho, The. See:
LA PAZ, BOLIVIA

Peaceful, The. See: PACIFIC
OCEAN

PEAK. See: INDIANOLA
PEAK

Peak XV. See: EVEREST,
MOUNT

Peaks, The Mountain of Three.
See: TITANO, MOUNT

PEARL HARBOR. Inlet on S.
coast of Island of Oahu,
Hawaiian Islands, 6 mi.
W. of Honolulu

The Gibraltar of the Pa-
cific

Pearl of the Antibes, The.

See: CUBA

Pearl of the Antilles, The.
See: CUBA

Pearl of the Atlantic, The.
See: MADEIRA

Pearl of the Mediterranean,
The. See: ALEXANDRIA,
EGYPT

Pearl of the Orient, The. See:
MANILA, P.I.

PEI-HO. River of Hopeh province, N.E. China, flowing
into Gulf of Po Hai at Taku;
about 350 mi. long

The White River

PEKING, CHINA. Capital city
of People's Republic of China sit. in N.E. China 527
mi. S.W. of Mukden

The Celestial City
The City of the Great Khan
The Forbidden City
The Northern Capital

Pelican Island. See: ALCA-
TRAZ ISLAND

Pelicans, The Island of. See:
ALCATRAZ ISLAND

PENANG, MALAYSIA. Seaport
and capital of state of Penang, Federation of Malaysia,
sit. on Prince of Wales Island in Straits of Malacca

George Town

PENINSULA. See: BOOTHIA
PENINSULA, CANADA; MA-
LAY PENINSULA (MALAY-
SIA); SALENTINE PENINSU-
LA

Penitentiary. See: Cradle of

the Penitentiary, The

People's Government, The
Home of the. See: TER-
IOKI, U.S.S.R.

PERNAMBUCO, BRAZIL.
Coastal city of Pernambuco
state, E. Brazil; now called
Recife

The City of the Reef

PERSEPOLIS, PERSIA. Cap-
ital of ancient Persia sit.
on plain of Merdusht, 35
mi. N.E. of Shiraz

The Glory of the East
The Throne of Jamsheed

PERSIAN GULF. Arm of
Arabian Sea between peninsula of Arabia and Iran

The Green Sea
Oman's Sea

Perspectives, The City of.
See: LENINGRAD, U.S.S.R.

PERTH, SCOTLAND. Burgh of
Perth County, central Scot-
land, sit. on Tay R. 32 mi.
N.W. of Edinburgh

The Fair City

Peru, The Treasury of. See:
ANDES MOUNTAINS

PETER, ST. See: COLLE-
GIATE CHURCH OF ST.
PETER

PETERBOROUGH, CANADA.
City of S.E. Ontario, Canada, sit. on Otonabee R.
13 mi. N. of Rice Lake

The Home of the Peter-
borough Canoe
The Home of the Rice

Lake Canoe

PETRA, ARABIA. Ancient
city of N. Arabia located in
Desert of Edom, 18 mi.
N.W. of Ma'an

The Enchanted City

Petrified City, The. See:
ISHMONIE, EGYPT

Pharaoh, The Treasury of.
See: EL KHAZNA

Pharaohs, The Land of the.
See: EGYPT

Philadelphia, The Long Branch
of. See: MAY, CAPE

PHILADELPHIA, PALESTINE.
Ancient city of E. Palestine
sit. near site of modern
Amman, 25 mi. N.E. of
Dead Sea

The Chief City of the Am-
monites

PHTHIA, GREECE. Ancient
district of S. Thessally,
N.E. Greece, now part of
Phthiotis and Phocis dept.

The Residence of Achilles

Pied Piper, The Town of the.
See: HAMELN

Pilgrim Port for Jerusalem,
The. See: JAFFA, ISRAEL

Pillar River. See: CONGO
RIVER

Pillars of Hercules, The. The
Gates of Hercules. Names
given by ancients to two
peaked rocks sit. at Gibraltar, which connects Med-
iterranean Sea and Atlan-
tic Ocean at S. tip of Iberian Peninsula

PILSEN. See: PLZEN (PIL-
SEN)

Pilsener Beer, The Home of.
See: PLZEN (PILSEN)

Piper, The Town of the Pied.
See: HAMELN

Pirate Port, The. See: PORT
ROYAL, JAMAICA

Pirate's Home, The. See:
BARATARIA BAY

Pirates' Stronghold, The Barbary. See: TRIPOLI,
LIBYA

PISA, ITALY. City and capital of Pisa province, Tuscany, Italy, 43 mi. S.W. of
Florence

The City of the Leaning
Tower

Pistol, The Birthplace of the.
See: PISTOLA, ITALY

PISTOLA, ITALY. Capital city
of province of same name,
Tuscany, Italy, 21 mi. N.W.
of Florence

The Birthplace of the Pistol

Pit. See: Money Pit, The

PITCAIRN ISLAND. Volcanic
island of S. Pacific Ocean
lying about midway between
Australia and S. America
and 100 mi. S.E. of Tuamotu
Archipelago

The "Bounty" Island
The Island of Mutineers

Pittsburgh of China, The.
See: HANKOW, CHINA

Pittsburgh of France, The.

See: SAINT-ETIENNE,
FRANCE

Pittsburgh of Japan, The. See:
YAWATA, JAPAN

Place of Martyrdom, The.
See: MESHED, IRAN

Place of Meeting, A. See:
TORONTO, CANADA

Place of the Big Three Conference, The. See: YAL-
TA, RUSSIA

Place of the Druids, The. See:
ANGLESEY, ENGLAND

Place of the Great Temple,
The. See: KARNAK, UNIT-
ED ARAB REPUBLIC

Place of the Hermits, The.
See: EINSIEDELN, SWITZ-
ERLAND

Place of the Hindus, The. See:
HINDUSTAN

Place of the Muddy Waters,
The. See: SHABARAKH
USU

Place of the Royal Oak, The.
See: BOSCOBEL, ENGLAND

Place of Weeping, The. See:
WEENEN, SOUTH AFRICA

Places, The Queen of English
Watering. See: SCARBOR-
OUGH, ENGLAND

Plain. See: Eurasian Plain,
The

Plain, The Cities of the. See:
GOMORRAH; SODOM

Plains. See: Great Plains, The

PLAINS OF ABRAHAM. Pla-

teau W. of Quebec, Canada;
battlefield (September 13,
1759) where British under
General Wolfe defeated
French under General Montcalm

Montcalm's Waterloo

Plant, The Home of the Coffee.
See: KAFA (KAFFA)

Plateau. See: Telegraph Plateau, The

Play, The Home of the Passion.
See: OBERAMMERGAU, BA-
VARIA

Playground of Europe, The.
See: SWITZERLAND

Playground of Two Nations,
The. See: GREAT SMOKY
MOUNTAINS

Pleasure, The London of Fashion and. See: WEST END

Plenty, The Land of. See:
ITALY

PLYMOUTH COLONY. Colony
established in S.E. Massachusetts by Pilgrims in 1620

The Old Colony

PLZEN (PILSEN), CZECHO-
SLOVAKIA. Capital of W.
Bohemia region, Czecho-
slovakia, 52 mi. W.S.W.
of Prague

The Home of Pilsener Beer

Podunk. Name used to designate any very small and in-
significant place. Hamlets
of this name are located in
states of Massachusetts and
Connecticut

POINT. See: STEEP POINT

Point, The. See: Loop, The

Point of Africa, The Westernmost. See: ALMADIES,
CAPE

Point of Europe, The Westernmost. See: SAINT VIN-
CENT, CAPE

Point of the Air, The West.
See: RANDOLPH FIELD,
TEXAS

Point of the Burma Road, The
Starting. See: LASHIO,
BURMA

Point of the Reformation, The
Focal. See: WITTENBERG,
GERMANY

Poland. See: Versailles of
Poland, The

POLE. See: NORTH POLE;
SOUTH POLE

Pole, The Home of the North
Magnetic. See: BOOTHIA
PENINSULA, CANADA

Pompeii, The Numidian. See:
TIMGAD, ALGERIA

Pond. See: Bloody Pond, The

Pond, The Herring. See: AT-
LANTIC OCEAN

Ponies, The Islands of. See:
SHETLAND ISLANDS

Pool, The. Part of Thames
R., England, immediately
below London Bridge

Pool of Immortality, The City
of the. See: AMRITSAR,
INDIA

Pope, The Home of the. See:
VATICAN CITY

POPOCATEPETL. Dormant
volcano in Puebla state,
Mexico, 40 mi. S. E. of
Mexico City

The Smoking Mountain

Population of the United States,
The Capital of the Negro.
See: HARLEM

Port, The (El Puerto). See:
PUERTO DE SANTA MA-
RIA, SPAIN

Port, The Pirate. See: PORT
ROYAL, JAMAICA

Port for Jerusalem, The Pilgrim. See: JAFFA, IS-
RAEL

Port of Five Seas, The. See:
MOSCOW, RUSSIA

Port of the Sudan in the Sahara, The. See: TIMBUK-
TU, AFRICA

PORT ROYAL, JAMAICA.
Fortified town at entrance
to Kingston Harbor, S. E.
Jamaica, British W. Indies;
destroyed by earthquake in
1692

The Pirate Port
The Sunken City

Port Wine, The Home of. See:
OPORTO, PORTUGAL

PORTO BELLO, PANAMA.
Seaport village on Carib-
bean coast of Panama sit.
20 mi. N. E. of Colón

The Sixteenth-Century Em-
porium of South Ameri-
can Trade

PORTPATRICK, SCOTLAND.
Decayed seaport, now sum-
mer resort of Wigtown county, S. W. Scotland; nearest
port of Great Britain to Ireland

The Gretna Green of Ireland

Ports, The Chief of the Cinque.
See: DOVER, ENGLAND

Portsmouth of the Steppes, The.
See: BAKU, U. S. S. R.

Portugal, The Granary of. See:
ALENTEJO, PORTUGAL

Portuguese Paradise, The. See:
SINTRA, PORTUGAL

Poseidon, The City of. See:
HELICE, GREECE

Potato, The Land of the. See:
IRELAND (EIRE)

Potato Land. See: IRELAND
(EIRE)

POTOMAC RIVER. River of
W. Virginia, Virginia and
Maryland, flowing from W.
Virginia to Chesapeake Bay;
about 400 mi. long

The River of Swans

Potteries, The Mother of the.
See: BURSLEM, ENGLAND

Pounded Corn, The River of
Coarse. See: CHICKAHOM-
INY RIVER

Powder Keg of Europe, The.
Collective appellation of
Balkan countries as a con-
sequence of the Balkan Wars
and World War I

PRE. See: GRAND PRE, NO-
VA SCOTIA

Present Meet, Where Past and.

See: CAMBODIA

President's Retreat, The. See:
CAMP DAVID, MARYLAND

Prester John, The Land of.
See: ETHIOPIA (ABYS-
SINIA)

PRIBLOFF ISLANDS. Group
of islands sit. in Bering
Sea about 200 mi. N. of
Aleutian Islands

The Fur Seal Islands

Print, The City of Jouy. See:
JOUY-EN-JOSAS, FRANCE

Printing, The Birthplace of.
See: MAINZ, GERMANY

PRISON. See: ANDERSON-
VILLE PRISON; NEWGATE
PRISON

Prison, America's First Escape-Proof. See: ALCA-
TRAZ ISLAND

Prison, Newgate. See: Birdcage Walk

Prisoner, The Home of the.
See: CHILLON

Producing State, The Beer.
See: BAVARIA

Profile Mountain, The. See:
CANNON MOUNTAIN

Promise, The Island of. See:
PUERTO RICO

Promise, The Land of. See:
CANAAN, PALESTINE

Promised Land, The. See:
CANAAN, PALESTINE

PROMONTORY. See: WIL-
SON'S PROMONTORY

Promontory, The Sacred. See:
SAINT VINCENT, CAPE

Proof Prison, America's First
Escape. See: ALCATRAZ
ISLAND

Prophet, The City of the. See:
MEDINA, SAUDI ARABIA

Prophets, The Land of Kings
and. See: PALESTINE

Prophet's City, The. See:
MEDINA, SAUDI ARABIA

Protection Wall, The Anti-
Fascist. See: BERLIN
WALL

Protestant Europe, The Intellectual Center of. See:
GENEVA, SWITZERLAND

Protestantism, The Rome of.
See: GENEVA, SWITZER-
LAND

Provençal Venice, The. See:
MARTIGUES, FRANCE

Province, The Fourth Part of
a. See: TETRARCHY OF
LYSANIAS

Province, The Model. See:
SHANSI, CHINA

Prussia, The Manchester of.
See: ELBERFELD, GER-
MANY

Puerto, El (The Port). See:
PUERTO DE SANTA MARIA,
SPAIN

PUERTO DE SANTA MARIA,
SPAIN. Commune of Cadiz
province, S.W. Spain, sit.
on Bay of Cadiz 8 mi. N.E.
of city of Cadiz

El Puerto (The Port)

PUERTO RICO.   Island of W.
Indies sit. in Atlantic Ocean
70 mi. E. of Hispaniola

The Island of Promise

Pulpit.   See: Hugh Lloyd's
Pulpit

PUNJAB, INDIA.   Region in
N.W. portion of subcontinent
of India, now partitioned into Union of India and Pakistan

The Empire of the Seven
Rivers
Five Rivers

Purchase. See: Gadsden Purchase, The

Pure, The Land of The.   See:
PAKISTAN

PURI, INDIA.   Seaport of
state of Orissa province,
Republic of India, 260 mi.
S.S.W. of Calcutta

The City of Jagannath

Purple Apennines, The.   See:
APENNINES

Purse of Spain, The Gold.   See:
ANDALUSIA, SPAIN

PUSHKIN, RUSSIA.   Town of
N.W. Leningrad region, So-
viet Russia, Europe, 15 mi.
S. of Leningrad

The Children's Village
The Tsar's Village

Putsch, The Town of the Beer
Hall. See: MUNICH, BA-
VARIA

PYHÄKOSKI.   Rapids in Oulu
R. near its mouth, S.E. of
Oulu, W. Finland

The Holy Rapids

Pyramids, The City of.   See:
GIZA, EGYPT

Pyrrhic Victory, The City of.
See: HERACLEA, ITALY

-Q-

QASR. See: EL QASR, EGYPT

Quad. See: Tom Quad

Quadrant. See: Enderby Quadrant, The

Quarter. See: French Quarter,
The

QUEBEC, CANADA.   City of
Quebec province, Canada,
sit. on N. bank of St. Lawrence R. 180 mi. from Montreal

The Gibraltar of America

Queen-Faithful Isle, The. See:
CUBA

Queen of Cities, The. See:
ROME, ITALY

Queen of English Watering
Places, The. See: SCAR-
BOROUGH, ENGLAND

Queen of Roads, The. See:
APPIAN WAY

Queen of Spas, The. See:
BATH, ENGLAND

Queen of the Adriatic, The.
See: VENICE, ITALY

Queen of the Black Sea, The.
See: ODESSA, RUSSIA

Queen of the East, The.   See:

ANTIOCH, ASIA MINOR;
BATAVIA, NETHERLAND
INDIES

Queen of the Eastern Archipelago, The. See: JAVA

Queen of the Hanse, The. See:
LÜBECK, GERMANY

Queen of the Lakes, The. See:
WINDERMERE

Queen of the North, The. See:
EDINBURGH, SCOTLAND

Queen of the Sea, The. See:
TYRE, LEBANON

Queen of the World, The. See:
MEROË

Queen's Chamber, The. See:
SPITHEAD, ENGLAND

Queensland, The Garden City of.
See: TOOWOOMBA, AUS-
TRALIA

Quetzal, The Land of the. See:
GUATEMALA

-R-

Ra, The City of. See: HEL-
IOPOLIS, EGYPT

Races, The Home of the Castlebar. See: CASTLEBAR,
IRELAND

Rack, The Coat. See: SYD-
NEY HARBOR BRIDGE

Rammekens, The Netherlands.
See: Cautionary Towns,
The

RAMOTH GILEAD, PALES-
TINE. Ancient town of
Gilead, sit. E. of Jordan

and N. of Jabbock R.'s

The City of Refuge

Rampart, The Long. See:
GREAT WALL, THE

RAMSAY ABBEY. Ruined Ben-
edictine abbey sit. in Hunt-
ingdonshire, E. central England, 65 mi. N. of London

The Croesus of English Abbeys
Ramsay the Rich

Rand, The. See: WITWATERS-
RAND

RANDOLPH FIELD, TEXAS.
U.S. Air Force base and
military reservation sit.
18 mi. N.E. of San An-
tonio

The West Point of the Air

Rapids, The Holy. See:
PYHÄKOSKI

Realm, Montezuma's. See:
MEXICO

RECIFE, BRAZIL. Capital
city of state of Pernambuco,
Brazil, sit. at mouth of
Capibaribe R. on Atlantic
Ocean

The Venice of America

Red Basin, The. See: SZECH-
WAN, CHINA

RED SEA. Inland sea extend-
ing N.N.W. from Strait of
Bab el Mandeb to Suez,
Egypt, separating coasts of
Arabia and N.E. Africa

The Sea of Sedge

Reef, The City of the. See:

PERNAMBUCO, BRAZIL

Reekie, Auld. See: EDIN-
BURGH, SCOTLAND

Reformation, The Cradle of
the. See: WITTENBERG,
GERMANY

Reformation, The Focal Point
of the. See: WITTENBERG,
GERMANY

Reformation, The Home of the.
See: WITTENBERG, GER-
MANY

Refuge, The City of. See:
RAMOTH GILEAD, PAL-
ESTINE

Refuge, The City of the. See:
MEDINA, SAUDI ARABIA

REINET. See: GRAAF REI-
NET, SOUTH AFRICA

Religious City, The. See:
ATHENS, GREECE

Remote Headstream of the
Nile, The Most. See:
KAGERA

REPUBLIC. See: CONGO,
REPUBLIC OF THE; DOM-
INICAN REPUBLIC

Republic. See: Mountain Republic, The

Republic, The Black. See:
DOMINICAN REPUBLIC;
HAITI

Republic, The Helvetian. See:
SWITZERLAND

Republic, The Shoestring.
See: CHILE

Republic in the World, The
Smallest. See: SAN MA-
RINO

Republic of China, The Home
of the. See: TAIWAN
(FORMOSA)

Republic of Free Men, The.
See: TAIWAN (FORMOSA)

Residence of Achilles, The.
See: PHTHIA, GREECE

Resort, The Millionaire's.
See: JEKYLL ISLAND

Resort City, The Egyptian.
See: ASWAN, EGYPT

Resplendent, The. See: SRI
LANKA (CEYLON)

REST ISLAND, MINN. Lake
sit. on W. side of Lake
Pepin, Minnesota

The Little Switzerland of
America

Retreat, The President's. See:
CAMP DAVID, MARYLAND

Return, The Isle of No. See:
ALCATRAZ ISLAND

Return, The River of No. See:
BIG SALMON RIVER

Rhine Cities. See: League of
Rhine Cities, The

Rhine of America, The. See:
HUDSON RIVER

Rhodesian Man, The Home of
the. See: BROKEN HILL,
AFRICA

Rice Bowl, The. Appellation
of E. central China pro-
vinces of Hunan and Hupeh,
where rice is a principal
crop

Rice Bowl of the World, The.
See: BANGKOK, THAI-
LAND

Rice Lake Canoe, The Home
of the.   See: PETERBOR-
OUGH, CANADA

Rich, Ramsay the.   See: RAM-
SAY ABBEY

RICO.   See: PUERTO RICO

Ride, The Town of Godiva's.
See: COVENTRY, ENG-
LAND

Ridge, Bloody Nose.   See:
UMURBROGOL

Ringing Island, The.   See:
ENGLAND

RIO DE JANEIRO, BRAZIL.
Commercial seaport of
Brazil, sit. on S.W. shore
of Guanabara Bay, arm of
Atlantic Ocean

Río

RIO GRANDE.   River of N.
America flowing from San
Juan Mts. in W. Colorado
to Gulf of Mexico; marks
part of border between
U.S. and Mexico; about
1800 mi. long

The River of the Wetbacks

RIO ROOSEVELT.   River sit.
in W. central Brazil, flowing from W. Mato Grosso
state to Aripuaña R.; about
200 mi. long

The River of Doubt

Rising Sun, The Coast of the.
See: RIVIERA, THE

Rising Sun, The Land of the.
See: JAPAN

Rising Yen, The Land of the.
See: JAPAN

Ritchie's Archipelago.   Series
of small islands E. of main
group of Andaman Islands
in Bay of Bengal

RIVER.   See: AMAZON RIV-
ER; AMERICAN RIVER;
BIG SALMON RIVER; BRAH-
MAPUTRA RIVER; CHICA-
GO RIVER; CHICKAHOMINY
RIVER; COLUMBIA RIVER;
CONGO RIVER; DANUBE
RIVER; DETROIT RIVER;
GANGES RIVER; GUADAL-
QUIVIR RIVER; HUDSON
RIVER; IRRAWADDY RIVER;
JORDAN RIVER; LIMPOPU
RIVER; MAHANADI RIVER;
MISSISSIPPI RIVER; MIS-
SOURI RIVER; NILE RIVER;
POTOMAC RIVER; SAINT
JOHN'S RIVER; TIBER RIV-
ER; VOLGA RIVER; WEI
RIVER VALLEY; WISCON-
SIN RIVER; YALU RIVER;
YANGTZE KIANG RIVER

River, Big Muddy.   See: MIS-
SOURI RIVER

River, The Golden Sand.   See:
KINSHA

River, The Great.   See: GUA-
DALQUIVIR RIVER; MA-
HANADI RIVER; MISSISSIPPI
RIVER; NILE RIVER

River, The Holy.   See: GAN-
GES RIVER

River, The Imperial.   See:
GRAND CANAL

River, The Infant Mississippi.
See: ITASCA, LAKE

River, The Mouth of the Han.
See: HANKOW, CHINA

River, Nicollet's Infant Mississippi.   See: ITASCA,
LAKE

River, Old Man. See: MIS-
SISSIPPI RIVER

River, Pillar. See: CONGO
RIVER

River, The Sacred. See:
GANGES RIVER

River, The Transport. See:
GRAND CANAL

River, West. See: SI-KIANG

River, The White. See: PEI-
HO

River, The Yang Kingdom.
See: YANGTZE KIANG
RIVER

River, The Yellow. See:
HWANG HO

River of a Thousand Isles,
The. See: WISCONSIN
RIVER

River of Coarse Pounded Corn,
The. See: CHICKAHOMINY
RIVER

River of Crocodiles, The. See:
LIMPOPU RIVER

River of Doubt, The. See:
RIO ROOSEVELT

River of Gold, The. See:
AMERICAN RIVER

River of No Return, The.
See: BIG SALMON RIVER

River of Steep Hills, The.
See: HUDSON RIVER

River of Swans, The. See:
POTOMAC RIVER

River of the West, The Hudson. See: MISSISSIPPI
RIVER

River of the Wetbacks, The.
See: RIO GRANDE

River Styx, The. See: MAV-
RO NERO

River That Runs Backward,
The. See: CHICAGO RIV-
ER

Rivers, The Achilles of. See:
COLUMBIA RIVER

Rivers, The Country Between
Two. See: MESOPOTAMIA

Rivers, The Empire of the
Seven. See: PUNJAB,
INDIA

Rivers, Five. See: PUNJAB,
INDIA

RIVIERA, THE. Portion of
European Mediterranean
coast extending from Hyères,
Var department, France, to
La Spezia, Ligurian region,
Italy

The Azure Coast
The Coast of the Rising Sun
The Coast of the Setting Sun

Riviera, The Center of the
Soviet. See: YALTA, RUS-
SIA

ROAD. See: WILDERNESS
ROAD, THE

Road, The Starting Point of
the Burma. See: LASHIO,
BURMA

Roads, The Queen of. See:
APPIAN WAY

Roaring Forties, The. Stormy
belt in N. Atlantic Ocean
between 40th and 50th paral-
lels of N. latitude

Robin Hood, The Home of.

See: SHERWOOD FOREST

Robin Hood's Forest. See: SHERWOOD FOREST

ROCHDALE, ENGLAND. Borough of Lancashire, England, 196 mi. N.W. of London

The Birthplace of the Cooperative Movement

ROCK. See: DUMBARTON, ROCK OF; INCHCAPE ROCK

Rock. See: Dome of the Rock, The

Rock, The. See: ALCATRAZ ISLAND; CORREGIDOR ISLAND; GIBRALTAR

Rock, The Bell. See: INCHCAPE ROCK

Rock, The Dragon's. See: DRACHENFELS

Rock, The Ladies'. See: STIRLING, SCOTLAND

Rock of Corregidor, The. See: CORREGIDOR ISLAND

Rock of Gibraltar, The. See: GIBRALTAR

Rock of Italy, The Great. See: CORNO, MONTE

Rocket Coast, The. Area of France, including Arras, about 120 mi. N.E. of Paris; heavily bombed by Allies during World War II

Rockies, The. See: ROCKY MOUNTAINS

ROCKY MOUNTAINS. Chain of mountain ranges in North America extending from N. Mexico to N. Alaska

The Backbone of North America
The Backbone of the American Continent
The Rockies
The Roof of the Continent

ROMAN EMPIRE. See: HOLY ROMAN EMPIRE

Roman Empire, The. See: HOLY ROMAN EMPIRE

Roman World, The Granary of the. See: EGYPT

ROME, ITALY. Capital and largest city of Italy; one of the oldest cities in the world

The City of Palaces
The City of Seven Hills
The Eternal City
The Hill City
The Holy City
The Imperial City
The Mistress of the World
The Nameless City
The Niobe of Nations
The Queen of Cities
The Seven-Hilled City

Rome, The Beautiful Daughter of. See: FLORENCE, ITALY

Rome, The Second. See: AQUILEIA, ITALY

Rome of Buddhism, The. See: LHASA (LASSA), TIBET

Rome of Hindustan, The. See: AGRA, INDIA

Rome of Protestantism, The. See: GENEVA, SWITZERLAND

Rome of the North, The. See:

COLOGNE, GERMANY

Roof of the Continent, The.
See: ROCKY MOUNTAINS

Roof of the World, The. See:
PAMIR

Rooftop of Eastern America,
The. See: GREAT SMOKY
MOUNTAINS

ROOSEVELT. See: RIO
ROOSEVELT

ROQUEFORT-SUR-SOULZON,
FRANCE. Town of S.E.
Aveyron dept., France

The Town of Cheese

Rose, The Land of the. See:
ENGLAND

Roses, The City of. See:
LUCKNOW, INDIA

Rotten Row. Fashionable
thoroughfare for equestrians
in Hyde Park, London,
England

ROUEN, FRANCE. Capital of
department of Seine-Mari-
time, France; former cap-
ital of ancient province of
Normandy, 87 mi. N.W. of
Paris

The Manchester of France

Route. See: Great Ox-Bow
Route, The

Row. See: Millionaire's
Row; Rotten Row

ROYAL. See: PORT ROYAL,
JAMAICA

Royal City of Herod, The
Once. See: TIBERIAS,
PALESTINE

Royal Oak, The Place of the.
See: BOSCOBEL, ENGLAND

Royal Town of Cunobelin, The.
See: COLCHESTER, ENG-
LAND

Ruby, The Land of the Hyacinth and. See: SRI-LANKA

RÜGEN, GERMANY. Largest
of the islands of Germany,
sit. in Baltic Sea opposite
Pomeranian Coast

The Holy Island

RUNNYMEDE, ENGLAND.
Meadow on Thames R. in
Surrey, England, where
King John signed the Magna
Charta on June 15, 1215

The Birthplace of English
Justice
The Birthplace of the Mag-
na Charta

Runs Backward, The River
That. See: CHICAGO RIV-
ER

RUSSIA. Empire in E. Europe
and N.W. Asia

The Northern Bear
The Northern Giant

Russia, The Birmingham of.
See: TULA, RUSSIA

Russia, The Key of. See:
SMOLENSK, U.S.S.R.

Russian Ireland. Appellation
of the Baltic provinces of
Russia

Russians, The Holy Mother of
the. See: MOSCOW, RUS-
SIA

RUWENZORI, MOUNT. Mountain group in central Africa, sit. between Lake Albert and Lake Edward

The Mountains of the Moon

-S-

SABA. Ancient country in S. Arabia, probably including Yemen and Hadhramaut

The Land of Myrrh

Sabines, The Home of the. See: CURES, ITALY

SABLE, CAPE. Cape at S.W. tip of peninsula of Florida enclosing Whitewater Bay; southernmost point of U.S.

America's Farthest South

Sacred City, The. See: JER-USALEM, PALESTINE

Sacred Island, The. See: IRELAND (EIRE)

Sacred Mountain, The. See: FUJI (FUJIYAMA), MOUNT

Sacred Mountain, China's. See: TAI-SHAN

Sacred Promontory, The. See: SAINT VINCENT, CAPE

Sacred River, The. See: GAN-GES RIVER

SAFED KOH. Mountain range of Afghanistan; a W. continuation of the Koh-i-Baba

White Mountain

SAFETY ISLANDS (ISLES DU SALUT). Group of three islands off coast of French

Guiana, S. America, opposite mouth of Maroni R.; part of French penal settlement

Devil's Islands (Isles du Diable)

Sagamore Hill. Home of Theodore Roosevelt sit. in village of Cove Neck near town of Oyster Bay, on Long Island Sound, N.Y.

Sahara, The Port of the Sudan in the. See: TIMBUKTU, AFRICA

Sailors' Hell, The. See: SHANGHAI, CHINA

Sailors' Snug Harbor. See: STATEN ISLAND

SAIMAA, LAKE. Lake in S.E. Finland, sit. in Viipuri and Mikkeli depts.

The Lake of the Thousand Lakes

SAINT ANDREWS. Royal burgh of Fifeshire, E. Scotland, 15 mi. S.E. of Dundee

The Golfer's Paradise

St. Bernard, The Home of. See: CLAIRVAUX, FRANCE

SAINT-ETIENNE, FRANCE. Manufacturing city of Loire dept., S.E. central France, 32 mi. S.W. of Lyons; site of many steel mills

The Pittsburgh of France

St. George, The City of. See: LOD, PALESTINE

SAINT GILES'S CHURCH. Sixteenth-century church sit. in

Wrexham, Denbighshire, N.
Wales

One of the Seven Wonders
of Wales

SAINT JOHN'S, NEWFOUND-
LAND.   Capital city of New-
foundland, 560 mi. N. E. of
Halifax; nearest port in N.
America to Europe

The Gibraltar of North
America

SAINT JOHN'S RIVER.   River
flowing from Lake Helen
Blazes through N. E. Florida
to Atlantic Ocean; about
350 mi. long

The American Nile

ST. PETER. See: COLLE-
GIATE CHURCH OF ST.
PETER

SAINT VINCENT, CAPE.
Cape sit. at S. W. point
of Portugal, about 118 mi.
S. of Lisbon

The Sacred Promontory
The Westernmost Point of
Europe

Saints, Aran of the. See:
ARAN ISLANDS

Saints, The Island of the.
See: IRELAND (EIRE)

Saints, The Isle of. See:
IRELAND (EIRE)

SALENTINE PENINSULA.
Peninsula of S. Italy sit.
between Adriatic Sea and
Gulf of Taranto

The Heel of Italy

SALMON RIVER. See:

BIG SALMON RIVER

Salt Desert, The Great. See:
DASHT-I-KAVIR

Salt Pan Hill. See: ZOUT-
PANSBERG

Salt Sea, The. See: DEAD
SEA

Salts, The Home of Epsom.
See: EPSOM, ENGLAND

SALUT, ISLES DU. See:
SAFETY ISLANDS (ISLES
DU SALUT)

Salvador, The Lighthouse of
El. See: IZALCO

SALVADOR, SAN. See: SAN
SALVADOR (WATLINGS
ISLAND)

SAMOA ISLANDS.   Groups of
islands in Pacific Ocean
about 2250 mi. S. S. W. of
Honolulu

Navigators Islands

Samuel, The Home of Eli and.
See: SHILOH, PALESTINE

SAN CLEMENTE, CALIFOR-
NIA.   S. California residence
of Richard M. Nixon, former
President of the United States

The Former Western White
House
The Western White House

SAN JOAQUIN VALLEY.   Valley of central California
about 30 mi. wide and 200
mi. long

The Granary of California

SAN MARINO.   Republic of S.
Europe on Italian peninsula,

11 mi. S.S.W. of Rimini,
Italy; occupies 38 sq. mi.

The Oldest State in Europe
The Smallest Republic in
the World

SAN SALVADOR (WATLINGS
ISLAND). One of Bahama
Islands sit. E.S.E. of Cat
Island in Atlantic Ocean

Columbus's First Landfall

Sanctorum, Insula. See: IRE-
LAND (EIRE)

Sanctuary, Luther's. See:
JENA, GERMANY

Sanctuary of Freedom, The.
See: OLD SOUTH CHURCH

Sand River, The Golden. See:
KINSHA

Sandalwood Island. See: SUM-
BA ISLAND

SANDS. See: GOODWIN SANDS

SANDS STREET, BROOKLYN,
N.Y. Street of borough of
Brooklyn, N.Y.

The Barbary Coast of the
East

Sandy Island, The Great. See:
FRASER ISLAND

Sandy Waste, The. See: GOBI
DESERT

SANTA MARIA. See: PUERTO
DE SANTA MARIA, SPAIN

SANTIAGO DE COMPOSTELA.
Commune of La Coruña
province, N.W. Spain, sit.
32 mi. S.W. of La Coruña

The Mecca of Spain

SANTO DOMINGO, HISPANIOLA.
City sit. on S. coast of W.
Hispaniola, founded by Spanish in 1496; now called Ciudad Trujillo

The Oldest Settlement

SANTOS, BRAZIL. Seaport of
S.E. Sao Paulo State, S.E.
Brazil, 200 mi. W.S.W. of
Rio de Janeiro

The City of Coffee

SAP. See: TONLE SAP

Sappho, The Island of. See:
LESBOS

Sappho's Leap. See: DOUCATO,
CAPE

Saratoga, The European. See:
BADEN-BADEN, GERMANY

Saturnia. See: ITALY

Sauce, The Home of Worcestershire. See: WORCESTER,
ENGLAND

Saxon Manchester, The. See:
CHEMNITZ, GERMANY

Saxon Switzerland, The. Hilly
region S.E. of Dresden, Saxony, Germany, at E. end of
the Erz Gebrige

SCANDINAVIA. Peninsula of
Norway and Sweden; may also include Denmark and Iceland

The Land of the Midnight
Sun

SCARBOROUGH, ENGLAND.
Seaport and health resort
of Yorkshire, England, 37
mi. N.E. of York

The Queen of English Watering Places

SCHRECKHORN, GROSS.
Mountain peak in Bernese
Alps, S.W. central Switz-
erland, sit. between Fin-
steraarhorn and Wetterhorn

The Mountain of Terrors

SCHWARZE ELSTER. River
in central Germany flowing
from Saxony into Elbe R.;
about 125 mi. long

The Black Elster

SCHWYZ, SWITZERLAND.
Commune of E. central
Switzerland 22 mi. E. of
Lucerne

The Cradle of Swiss Free-
dom

Sciences, The Treasury of.
See: BUKHARA, ASIA

SCILLIUM, AFRICA. Ancient
town of Byzacium, Roman
province of Africa, sit. near
modern Tunisian town of
Sbeitla

The Town of Twelve Martyrs

Scotia. See: SCOTLAND

SCOTLAND. Geographical portion of the island of Great
Britain, N. of England

Caledonia
The Land of Cakes
The Land of the Leal
The Land of the Thistle
North Britain
Scotia

Scotland, The Garden of. See:
MORAYSHIRE

Scotland, The Gibraltar of.

See: DUMBARTON, ROCK
OF

SEA. See: ARAL, SEA OF;
AZOV, SEA OF; BALTIC
SEA; BERING SEA; BLACK
SEA; CARIBBEAN SEA;
DEAD SEA; GALILEE, SEA
OF; MEDITERRANEAN SEA;
RED SEA

Sea, The Bride of the. See:
VENICE, ITALY

Sea, The Fish. See: AZOV,
SEA OF

Sea, The Great. See: MEDI-
TERRANEAN SEA

Sea, The Green. See: PER-
SIAN GULF

Sea, The Hospitable. See:
BLACK SEA

Sea, The Inhospitable. See:
BLACK SEA

Sea, The Island. See: ARAL,
SEA OF

Sea, The Lion of the. See:
GOOD HOPE, CAPE OF

Sea, Oman's. See: PERSIAN
GULF

Sea, Our (Mare Nostrum). See:
MEDITERRANEAN SEA

Sea, The Queen of the. See:
TYRE, LEBANON

Sea, The Queen of the Black.
See: ODESSA, RUSSIA

Sea, The Salt. See: DEAD
SEA

Sea, The Silver City by the.
See: ABERDEEN, SCOT-
LAND

Sea, The South. See: PA-
CIFIC OCEAN

Sea, The Vermillion. See:
CALIFORNIA, GULF OF

Sea-Girt Isle, The. See:
ENGLAND

Sea of Chinnereth, The. See:
GALILEE, SEA OF

Sea of Cortes, The. See:
CALIFORNIA, GULF OF

Sea of Darkness, The. See:
ATLANTIC OCEAN

Sea of Kamchatka, The. See:
BERING SEA

Sea of Lot, The. See: DEAD
SEA

Sea of Sedge, The. See: RED
SEA

Sea of Stars, The. See:
HWANG HO

Sea of the New World, The.
See: CARIBBEAN SEA

Sea of Tiberias, The. See:
GALILEE, SEA OF

Seal Islands, The Fur. See:
PRIBLOFF ISLANDS

Search of a Future, The Country in. See: MALAY
ARCHIPELAGO

Seas, The Circuit of the Northern. See: HOKKAIDO, JA-
PAN

Seas, The Key of the Dutch.
See: FLUSHING, THE
NETHERLANDS

Seas, The Mistress of the.
See: GREAT BRITAIN

Seas, The Port of Five. See:
MOSCOW, RUSSIA

Seat. See: Earl's Seat, The

Seat of the Council of Trent,
The. See: TRENTO, ITALY

Seat of the Eleusinian Mysteries, The. See: ELEUSIS,
GREECE

Seat of the Mukden War Lord,
The. See: MUKDEN, MAN-
CHURIA

Second Rome, The. See:
AQUILEIA, ITALY

Sedge, The Sea of. See: RED
SEA

Selkirk's Island. See: MAS
A TIERRA

Setting Sun, The Coast of the.
See: RIVIERA, THE

Settlement, Columbus's First.
See: ISABELA, NORTH
DOMINICAN REPUBLIC

Settlement, The Oldest. See:
SANTO DOMINGO, HISPAN-
IOLA

SETTLEMENTS. See: STRAITS
SETTLEMENTS

Seven Cities, The. Collective
appellation of Jerusalem,
Babylon, Athens, Rome,
Constantinople, London and
Paris

Seven Cities of Cibola, The.
Legendary group of cities
supposedly located in Zuni
pueblo of New Mexico;
thought to be fabulously rich

Seven Communes, The. Dis-
trict in Vicenza, N. Italy

Seven Falls, The. See: GUA-
IRA

Seven-Hilled City, The. See:
ROME, ITALY

Seven Hills, The City of. See:
ROME, ITALY

Seven Hunters, The. See:
FLANNAN ISLANDS

Seven Isles of Izu, The. See:
IZU SHICHITO

Seven Rivers, The Empire of
the. See: PUNJAB, INDIA

Seven Wonders of the Ancient
World, The. Collective appellation of Egyptian pyramids, hanging gardens of
Babylon, temple of Diana at
Ephesus, statue of Zeus by
Phidias at Olympia, Mauso-
leum at Halicarnassus, Col-
ossus of Rhodes and the
Pharos (lighthouse) at Alexandria

Seven Wonders of the World,
The. Collective appellation
of the Coliseum at Rome,
catacombs of Alexandria,
Great Wall of China, "Dru-
idical" temple at Stonehenge,
England, leaning tower of
Pisa, porcelain tower of
Nanking and the mosque of
St. Sophia at Constantinople

Seven Wonders of Wales, One
of the. See: SAINT GILES'S
CHURCH

SHABARAKH USU. Archaeo-
logical site sit. in S. Outer
Mongolia about 600 mi.
N.W. of Wanchuan (Kalgan)

The Place of the Muddy
Waters

Shamrock, The Land of the.

See: IRELAND (EIRE)

SHAN. See: TAI-SHAN; TIEN-
SHAN; WAN-SHOU-SHAN

SHANGHAI, CHINA. Largest
city and seaport of People's
Republic of China, sit. on
Hwang Pu R., tidal inlet of
Yangtze Kiang R. estuary

The Sailors' Hell

SHANSI. See also: Shensi

SHANSI, CHINA. Province of
N.E. China; one of the Five
Northern Provinces

The Home of Early Chinese
Agriculture
The Model Province

Sheffield of Germany, The.
See: SOLINGEN, GERMANY

Sheikdom, The Oil. See:
LIBYA (LIBIA)

Shensi. See also: SHANSI

Shensi, The Gateway to. See:
TUNGKWAN, CHINA

SHERWOOD FOREST. Ancient
royal forest in W. Nottinghamshire, England

The Home of Robin Hood
Robin Hood's Forest

SHETLAND ISLANDS. Archi-
pelago in N. Atlantic Ocean
50 mi. N.N.E. of Orkney
Islands

The Islands of Ponies

SHICHITO. See: IZU SHICHITO

SHILLELAGH, IRELAND. Town
of Wicklow, Ireland, famous
for its oak trees

The Town of the Cudgel

SHILOH, PALESTINE. Ancient town of the tribe of Ephraim sit. 20 mi. N. of Jerusalem; first permanent resting place of the Tabernacle

The Home of Eli and Samuel

Ship-Swallower, The. See: GOODWIN SANDS

SHISHALDIN. Active volcano sit. on S. Unimak Island, S.W. Alaska

Smoking Moses

Shivering Mountain, The. See: MAM TOR

Shoestring District, The. Appellation of 3rd Congressional District, Mississippi

Shoestring Republic, The. See: CHILE

Shore. See: French Shore, The

SHOU-SHAN. See: WAN-SHOU-SHAN

SI-HU. Lake within city of Hangchow, Chekiang province, E. China

The Western Lake

SI-KIANG. River of S.W. China flowing from province of Yunnan near Manning Hein to China Sea; about 1650 mi. long

The Great Commercial Highway of Southeast China
West River

Siberia, The Chicago of. See: NOVOSIBIRSK, RUSSIA

Siberia, The Icebox of. See: OIMYAKON, RUSSIA

Sibyl, The Home of the. See: CUMAE, ITALY

SICILY. Largest island of Mediterranean Sea sit. S.W. of Italian peninsula

The Garden of Italy
The Granary of Europe
The Granary of Italy

Sicily, The Granary of. See: CATANIA, SICILY

Sick Man, The. See: TURKEY

Sick Man of Europe, The. See: TURKEY

Sick Man of the East, The. See: TURKEY

Side. See: East Side; West Side

SIERRA MADRE. Mountains sit. in S. California near Los Angeles

The Graham McNamee Mountains

Sigh. See: Last Sigh of the Moor, The

Sighs. See: Bridge of Sighs, The

Silent City, The. See: AMYCLAE, GREECE; VENICE, ITALY

Silent Sister, The. See: TRINITY COLLEGE

Silk, The City of. See:

TOURS, FRANCE

Silver, The City of. See:
SUCRE, BOLIVIA

Silver, The Mountain of. See:
DAVIDSON, MOUNT

Silver City by the Sea, The.
See: ABERDEEN, SCOT-
LAND

Silver Hills. See: TEGUCI-
GALPA, HONDURAS

Silver Streak, The. See:
ENGLISH CHANNEL

Silvio, Monte. See: MAT-
TERHORN, THE

SIMONS-TOWN, SOUTH AFRI-
CA. Town of S.W. Cape
province, Union of S. Africa, 20 mi. S. of Cape Town

The Gibraltar of the South

Simple, The City of the. See:
GHEEL, BELGIUM

Sinai, Mount. See: JEBEL
MUSA

SINGAPORE, ASIA. Seaport
city on S. coast of Singapore Island sit. on Singapore Strait

The Crossroads of the Far
East
The Light of the South
The Lion City

SINGAPORE ISLAND. Island
off S. end of Malay Peninsula, S.E. Asia; main part
of Singapore Crown Colony

The Citadel of the Orient
The Island Nation
The Vaunted Island

SINTRA, PORTUGAL. Com-

mune of W. Lisbon district,
W. Portugal, 12 mi. N.W.
of Lisbon

The Portuguese Paradise

Sister, The Silent. See:
TRINITY COLLEGE

Sister Isle, The. See: IRE-
LAND (EIRE)

Sisters. See: Wayward Sisters, The

Site of the Nativity, The. See:
BETHLEHEM, PALESTINE

Six Cities, The. Collective
appellation of German cities
of Bautzen, Görlitz, Kamenz,
Lauban, Löbau and Zittau

Six Counties, The. Collective
appellation of Northern Irish
counties of Antrim, Armagh,
Down, Fermanagh, London-
derry and Tyrone

Sixteenth-Century Emporium of
South American Trade, The.
See: PORTO BELLO, PAN-
AMA

Skiing Technique, The Home of
the Arlberg. See: ARL-
BERG, AUSTRIA

Skulls, The Kirk of. See:
GOWRIE CHURCH

Sky High Capital, The. See:
LA PAZ, BOLIVIA

Sleeping Giant, The. See:
CHINA

Sleeve, The (La Manche). See:
ENGLISH CHANNEL

Slot. See: South of the Slot

Slot, The. Long, open-water
passage in central Solomon

Islands, W. Pacific Ocean, running from Shortland Islands to Florida Island and Savo Island

SLOUGH, ENGLAND. Town of Buckinghamshire, England, 18 mi. W. of London; site of Herschel's observatory and telescope

The Home of the Telescope

Smallest Republic in the World, The. See: SAN MARINO

Smile, The Land of. See: THAILAND

Smiling Island, The. See: ISCHIA

Smoke, The. See: LONDON, ENGLAND

Smoke, The Big. See: LONDON, ENGLAND

Smoke, The City of. See: LONDON, ENGLAND

Smoke, The Great. See: LONDON, ENGLAND

Smokies, The. See: GREAT SMOKY MOUNTAINS

Smokies, The Great. See: GREAT SMOKY MOUNTAINS

Smoking Islands, The. See: KURIL (KURILE) ISLANDS

Smoking Moses. See: SHISHALDIN

Smoking Mountain, The. See: POPOCATEPETL

SMOLENSK, U.S.S.R. City of Russia, 244 mi. S.W. of Moscow; important railway junction of W. Soviet Union

The Key of Russia

SMYRNA, ASIA MINOR. Ancient city of Asia Minor sit. at head of Gulf of Izmir

The Crown of Ionia

Snow, The Abode of. See: HIMALAYA, THE

Snow, The City of. See: LENINGRAD, U.S.S.R.

Snow-Covered Mountain, The. See: ILLIMANI

Snow Mountain. See: HERMON, MOUNT

Snug Harbor, Sailors'. See: STATEN ISLAND

Sod, The Owld. See: IRELAND (EIRE)

SODOM. Ancient city in the Plain of Jordan; with Gomorrah, one of the Cities of the Plain

Solar City, The. See: BAALBEK, LEBANON

SOLINGEN, GERMANY. City of N. Rhine-Westphalia, W. Germany, 13 mi. E. of Düsseldorf; famous for cutlery manufactured there

The Sheffield of Germany

Solomon, The Throne of. See: TAKHT-SULAIMAN

SOMALILAND, AFRICA. Territory of E. Africa between equator and Gulf of Aden

The Horn of Africa

SOMNATH, W. INDIAN UNION.

Ancient port city on S. coast of Kathiawar, W. Indian Union, sit. near modern Veraval

The Home of the Gates of Somnath

Son of Brahma, The. See: BRAHMAPUTRA RIVER

Song, The Land of. See: ITALY

Sorrow, China's. See: HWANG HO

Sorrow, The Mountain of. See: EDINBURGH, SCOTLAND

SOULZON. See: ROQUEFORT-SUR-SOULZON, FRANCE

Source of the Mississippi, The. See: ITASCA, LAKE

South. See: Sunny South, The

South, America's Farthest. See: SABLE, CAPE

South, Australia's Farthest. See: WILSON'S PROMON-TORY

South, The Benares of the. See: CONJEEVERAM, INDIA

South, The Distant. See: VIET NAM

South, The Farthest. See: HOWE, CAPE

South, The Gibraltar of the. See: SIMONS-TOWN, SOUTH AFRICA

South, The Light of the. See: SINGAPORE, ASIA

South, Old. See: OLD SOUTH CHURCH

SOUTH AFRICA. Portion of African continent S. of middle course of Zambezi R.

The Great Thirst Land

SOUTH AFRICA. See also: AFRICA; EQUATORIAL AFRICA

South Africa, The Most English Town in. See: GRAHAMS-TOWN, SOUTH AFRICA

South America, The Archaeological Capital of. See: CUZCO, PERU

South America, The Athens of. See: BOGOTA, COLOMBIA

South America, The Heroin Crossroads of. See: PARAGUAY

South American Trade, The Sixteenth Century Emporium of. See: PORTO BELLO, PANAMA

SOUTH CHURCH. See: OLD SOUTH CHURCH

South of the Border. See: MEXICO

South of the Slot. Area in San Francisco, Calif., S. of California Street cable car slot

SOUTH POLE. S. extremity of earth's axis at 90° S. latitude

The Bottom of the World

South Sea, The. See: PACIFIC OCEAN

Southampton Fields. See: Field of the Forty Footsteps, The

Southeast China, The Great
Commercial Highway of.
See: SI-KIANG

Southern Capital, The. See:
NANKING, CHINA

Southern Land, The. See:
VIET NAM

Southern Ocean, The. See:
PACIFIC OCEAN

Soviet Manchester, The. See:
IVANOVO, RUSSIA

Soviet Riviera, The Center of
the. See: YALTA, RUSSIA

Spain, The Garden and Granary
of. See: ANDALUSIA,
SPAIN

Spain, The Garden of. See:
ANDALUSIA, SPAIN

Spain, The Gold Purse of.
See: ANDALUSIA, SPAIN

Spain, The Granary of. See:
ANDALUSIA, SPAIN

Spain, The Last Moorish
Stronghold in. See: GRA-
NADA, SPAIN

Spain, Little (La Española).
See: HISPANIOLA

Spain, The Mecca of. See:
SANTIAGO DE COMPOS-
TELA

Spanish Main, The. See:
CARIBBEAN SEA

Spanish Mark, The. Region
of N.E. Spain; boundary
area between Pyrenees and
Elbro R. established by
Charlemagne in A.D. 795

Spas, The Queen of. See:

BATH, ENGLAND

Sphinxland. See: EGYPT

Spice Islands, The. See:
MOLUCCAS

SPITHEAD, ENGLAND. Road-
stead on S. coast of England,
sit. between Portsmouth and
Isle of Wight; anchorage for
British navy

The Queen's Chamber

Springs, Warm. See: TIFLIS,
GEORGIA

Spudland. See: IRELAND
(EIRE)

SQUARE. See: WASHINGTON
SQUARE

Square, Bughouse. See: WASH-
INGTON SQUARE

SRI LANKA (CEYLON). Is-
land dominion in British
Commonwealth of Nations
sit. in Indian Ocean S. of
India

The Land of the Hyacinth
and Ruby
The Resplendent

STAGIRA, GREECE. Macedon-
ian town of N.E. Greece,
sit. on E. Chalcidice Peninsula on Strymonic Gulf

The Town of Aristotle

Stars, The Sea of. See:
HWANG HO

Stars and Stripes, The Land
of the. See: UNITED
STATES OF AMERICA

Starting Point of the Burma
Road, The. See: LASHIO,
BURMA

State, The Beer-Producing.
  See: BAVARIA

State in Europe, The Oldest.
  See: SAN MARINO

STATE STREET, BOSTON,
  MASS.   Street in financial
  district of Boston, Mass.

  The Wall Street of Boston

STATE STREET, CHICAGO,
  ILL.   Famous street of
  Chicago, Ill.

  That Great Street

STATEN ISLAND.   Island sit.
  in New York Bay 5 mi. S.
  of Manhattan

  Sailors' Snug Harbor

STATES.   See: INDIAN
  STATES; UNITED STATES
  OF AMERICA

States.   See: Papal States,
  The; Succession States, The

States, The.   See: UNITED
  STATES OF AMERICA

States, The Native.   See: IN-
  DIAN STATES

States, The United.   See:
  UNITED STATES OF AMER-
  ICA

States of the Church, The.   See:
  Papal States, The

Stationary Island Aircraft Carriers, The.   See: MARI-
  ANA ISLANDS

STAUBBACH.   Waterfall in S.
  part of Canton of Bern,
  Switzerland, 8 mi. S. of In-
  terlaken

  The Dust Stream

Steep Hills, The River of.
  See: HUDSON RIVER

STEEP POINT.   Extreme W.
  point of mainland of Australia sit. S. of Shark Bay

  Australia's Farthest West

STEINSÖY, NORWAY.   Nor-
  wegian island off Sogne
  Fjord; most westerly point
  of Norway

  Norway's Farthest West

Stem, The Main.   See: BROAD-
  WAY, NEW YORK CITY

Steppes, The Portsmouth of
  the.   See: BAKU, U.S.S.R.

STIRLING, SCOTLAND.   Capital and river port of Stir-
  lingshire, Scotland, sit. on
  Firth of Forth 35 mi. N.W.
  of Edinburgh

  The Key to the Highlands
  The Ladies' Rock

Stone, Cactus on a.   See:
  MEXICO CITY, MEXICO

Stone Face, The Great.   See:
  CANNON MOUNTAIN

Stone Money, The Island of.
  See: YAP

Stone of the Broken Treaty,
  The.   See: LIMERICK,
  IRELAND

Stoned, The White.   See:
  MOSCOW, RUSSIA

STONEHENGE, ENGLAND.
  Assemblage of stones on
  Salisbury Plain, 7 mi. N.
  of Salisbury, England; pos-
  sibly old Druidical temple
  dating back to Bronze Age

The Home of the Altar Stones

Stones, The Home of the Altar.
See: STONEHENGE, ENG-
LAND

Stones, The Valley of. See:
GUADALAJARA, SPAIN

Storks, The City of the. See:
TIMBUKTU, AFRICA

Storms, The Cape of. See:
GOOD HOPE, CAPE OF

Stormy Cape, The. See:
GOOD HOPE, CAPE OF

STRAIT. See: BASS STRAIT;
GIBRALTAR, STRAIT OF;
MALACCA, STRAIT OF; TOR-
RES STRAIT

Straits, The. Name used to
designate the Strait of Gibraltar, Strait of Malacca
and others, such as Bass
Strait, Torres Strait and
link between the Mediter-
ranean and Black Seas; also shortened designation of
Straits Settlements

STRAITS SETTLEMENTS.
Former British crown col-
ony on Strait of Malacca

The Straits

STRASBOURG, FRANCE. Cap-
ital city of department of
Bas-Rhin, N.E. France,
88 mi. N. of Basel, Switzerland

The City of Bells

STRATFORD-UPON-AVON,
ENGLAND. Municipal
borough of Worcester-
shire, England, 8 mi. S.
of Warwick; birthplace of
William Shakespeare

The Home of the Bard

Streak, The Silver. See: ENG-
LISH CHANNEL

Stream, The Dust. See:
STAUBBACH

STREET. See: GRISWOLD
STREET, DETROIT, MICH.;
LA SALLE STREET, CHI-
CAGO, ILL.; MARKET
STREET, SAN FRANCISCO,
CALIF.; MONTGOMERY
STREET, SAN FRANCISCO,
CALIF.; SANDS STREET,
BROOKLYN, N.Y.; STATE
STREET, BOSTON, MASS.;
STATE STREET, CHICAGO,
ILL.; WALL STREET, NEW
YORK CITY

Street. See: No. 10 Downing
Street

Street, The. Area in New
York City near and including Wall Street; financial
district

Street, That Great. See:
STATE STREET, CHICAGO,
ILL.

Street of America. See: Main
Street of America, The

Street of Boston, The Wall.
See: STATE STREET, BOS-
TON, MASS.

Street of Chicago, The Wall.
See: LA SALLE STREET,
CHICAGO, ILL.

Street of Detroit, The Wall.
See: GRISWOLD STREET,
DETROIT, MICH.

Street of Extremes, The. See:
WALL STREET, NEW YORK
CITY

Street of the West, The Wall.

See: MONTGOMERY
STREET, SAN FRANCIS-
CO, CALIF.

Strip. See: Cherokee Strip

Stripes, The Land of the Stars
and. See: UNITED STATES
OF AMERICA

Stronghold, The Barbary Pirates'. See: TRIPOLI,
LIBYA

Stronghold, The Castle. See:
EL QASR, EGYPT

Stronghold in Spain, The Last
Moorish. See: GRANADA,
SPAIN

Stronghold of Carpetani, The.
See: TOLEDO, SPAIN

Stronghold of Paganism, The.
See: HARZ MOUNTAINS

STRONSAY. One of the Ork-
ney Islands sit. in Atlantic
Ocean off N. coast of Scotland

The Isle of the Leper's Well

Styx, The River. See: MAV-
RO NERO

Submission, The Town of Humble. See: CANOSSA, IT-
ALY

Succession States, The. Collective appellation of central
European states of Czechoslovakia, Yugoslavia, Poland,
Romania, Austria and Hungary

SUCRE, BOLIVIA. Legal,
though nominal, capital of
Bolivia, sit. 318 mi. S.E.
of La Paz, administrative
center of that country

The City of Silver

SUDAN, AFRICA. Republic
of Africa, S. of Libya and
the United Arab Republic

The Country of the Blacks

Sudan in the Sahara, The Port
of the. See: TIMBUKTU,
AFRICA

SUEZ CANAL. Artificial wa-
terway extending across Isthmus of Suez between Med-
iterranean Sea and Gulf of
Suez

The Highway to India
The World Ditch

SUHL, GERMANY. Manufac-
turing city of Erfurt govt.
district, Saxony province,
Prussia, E. Germany, 30
mi. S.S.W. of Erfurt

The Armory of Germany

SULAIMAN. See: TAKHT-
SULAIMAN

Sulfur Island. See: IWO
JIMA

Sumatra, The Montenegro of.
See: ACHIN, N. SUMATRA

SUMBA ISLAND. Island in
Malay Archipelago, 40 mi.
S. of Flores; one of the
Sunda Islands

Sandalwood Island

Summit of the World, The.
See: EVEREST, MOUNT;
YELLOWSTONE NATIONAL
PARK

Sun, The City of the. See:
BAALBEK, LEBANON; CUZ-
CO, PERU; HELIOPOLIS,
EGYPT

Sun, The Coast of the Rising.

See: RIVIERA, THE

Sun, The Coast of the Setting.
See: RIVIERA, THE

Sun, The Gardens of the.
See: MALAY ARCHIPEL-
AGO

Sun, The House of the. See:
HALEAKALA

Sun, The Land of the. See:
HINDUSTAN

Sun, The Land of the Midnight.
See: SCANDINAVIA

Sun, The Land of the Rising.
See: JAPAN

Sunken City, The. See:
PORT ROYAL, JAMAICA

Sunken Continent, The. See:
ATLANTIS

Sunken Continent of the Pacific, The. See: LEMURIA

Sunken Island, The. See:
ATLANTIS

Sunny South, The. S. part of
United States of America,
S. of Mason-Dixon Line

Sunshine Coast, Australia's.
See: NOOSA HEADS

Superb, The. See: GENOA,
ITALY

SUPERIOR, LAKE. See:
Great Lakes, The

Sur, Mar del. See: PACIF-
IC OCEAN

SUR-SOULZON. See: ROQUE-
FORT-SUR-SOULZON,
FRANCE

SUTRO TUNNEL. Drainage

tunnel $4\frac{1}{2}$ mi. long con-
structed 1869-1878 for purpose of draining Comstock
Lode; sit. near Virginia
City, Nev.

The Great Gold Tunnel

Swallower, The Ship. See:
GOODWIN SANDS

Swamp, The. Region of lower
part of New York City

Swamp, The Lake of the Dismal. See: DRUMMOND,
LAKE

Swans, The River of. See:
POTOMAC RIVER

Sweden, The Garden of. See:
BLEKINGE, SWEDEN

Sweet Home House. See:
Home, Sweet Home House

Swiss Freedom, The Cradle of.
See: SCHWYZ, SWITZER-
LAND

SWITZERLAND. Federal re-
public of W. central Europe

The Helvetian Republic
The Playground of Europe

Switzerland. See: English
Switzerland, The; French
Switzerland, The; Saxon
Switzerland, The

Switzerland, The Athens of.
See: ZURICH, SWITZER-
LAND

Switzerland, The Loretto of.
See: EINSIEDELN, SWITZ-
ERLAND

Switzerland of America, The
Little. See: REST ISLAND,
MINN.

Switzerland of Central America, The. See: PANAMA

Switzerland of This Hemisphere, The. See: PANAMA

Sword Blades, The City of. See: TOLEDO, SPAIN

SYDNEY HARBOR BRIDGE. Steel-arch bridge sit. in Sydney, Australia; span of 1650 ft.; completed in 1931

The Coat Rack

Symbol of Modern Might, A. See: GIBRALTAR

Synod, The City of the. See: GANGRA, TURKEY

SZECHWAN, CHINA. Province of S. central China, S. of Kansu and Shensi provinces; largest province of China proper

The Red Basin

-T-

TAI-HU. Lake in Kaingsu and Chekiang provinces, E. China

The Great Lake

TAI-SHAN. Mountain of China sit. in W. Shantung province 32 mi. S. of Tsinan; 5048 ft. high

China's Sacred Mountain

Tailless Cat, The Isle of the. See: MAN, ISLE OF

Tailtean Games, The Home of the. See: TELLTOWN, IRELAND

TAIPEI, FORMOSA. Administrative center of Formosa, China, sit. in N. part of the island about 15 mi. inland from Keelung

The Hub of the Orient

TAIWAN (FORMOSA). Island in China Sea off Fukien province, S. E. China

The Home of the Republic of China
The Republic of Free Men
The Tight Little Island

TAKHT-SULAIMAN. Twin peaks at N. end of Sulaiman range, India, 11,100 ft. high

The Throne of Solomon

Talisman. See: Gate of the Talisman, The

TANEZROUFT. Desert region of S.W. Algeria and N. French Sudan; exceptionally barren region of Sahara Desert

The Desert Within a Desert

TANGIPAHOA. Parish of Louisiana; famous for strawberries grown there

Tangipahoa of the Crimson Carpet

Tank of the Golden Lilies, The. Quadrangle in temple sit. in city of Madura, S. Indian Union, 270 mi. S. S. W. of Madras

Tarik, The Hill of. See: GIBRALTAR

Tarnished Jewel, The. See:

ALCATRAZ ISLAND

Tarts, The Town of Banbury.
See: BANBURY, ENGLAND

TEAPOT DOME, WYOMING.
U.S. naval oil reservation
sit. in Natrona County, central Wyoming; illegally li-
censed to oil interests by
Secretary of Interior Albert
B. Fall in 1922

Fall's Fall

Tears. See: Villa of Tears,
The

Tears, The Gate of. See:
BAB EL MANDEB

Technique, The Home of the
Arlberg Skiing. See:
ARLBERG, AUSTRIA

TEGUCIGALPA, HONDURAS.
Capital of Republic of Honduras sit. on Choluteca R.
60 mi. N.E. of Gulf of
Fonesca

Silver Hills

TEL AVIV, ISRAEL.   Capital
of state of Israel

Miami Beach East

Telegraph Plateau, The.   Bottom of Atlantic Ocean be-
tween Newfoundland and Ireland where many submarine
cables have been laid

Telescope, The Home of the.
See: SLOUGH, ENGLAND

Tell, The Home of William.
See: ALTDORF, SWITZER-
LAND

TELLTOWN, IRELAND.   Vil-
lage of County Meath, Ire-

land, sit. 35 mi. N.W. of
Dublin

The Home of the Tailtean
Games

Temple, The City of the Golden.
See: AMRITSAR, INDIA

Temple, The Place of the
Great. See: KARNAK,
UNITED ARAB REPUBLIC

TEMPLE HILL.   In antiquity
easternmost hill of city of
Jerusalem, lying between
valleys of Kidron and Tyro-
poeon

The City of David
Zion

Temple of a Thousand Columns,
The.   Ruins of ancient temple located in seaport town
of Trincomalee, E. Sri Lanka, sit. on Bay of Bengal
110 mi. S.E. of Jaffa

Ten Thousand Ancients, The
Mountain of. See: WAN-
SHOU-SHAN

TENNESSEE VALLEY.   Area
in S. United States devel-
oped by Tennessee Valley
Authority in terms of flood
control, transportation and
generation of electricity

The Happy Valley

Tent, The. See: CAIRO,
EGYPT

TERIOKI, U.S.S.R.   Town of
N.W. Leningrad region,
U.S.S.R., sit. on N. shore
of Gulf of Finland 30 mi.
W.N.W. of Leningrad

The Home of the People's
Government

Territories. See: Neutral Territories, The

Terrors, The Mountain of. See: SCHRECKHORN, GROSS

TETRARCHY OF LYSANIAS. Ancient district of Roman Empire sit. in Abilene, S.W. Syria

    The Fourth Part of a Province

THAILAND. Kingdom of S. Asia sit. on Gulf of Siam, an arm of the China Sea

    The Flourishing Land of the Free
    The Land of Smile
    The Land of the White Elephant

That Great Street. See: STATE STREET, CHICAGO, ILL.

That Runs Backward, The River. See: CHICAGO RIVER

THEBES, EGYPT. Ancient city of Egypt sit. on both banks of Nile R. 480 mi. S. of present-day Cairo

    The City of Amon
    The City of Zeus
    The Hundred-Gated City

THERMOPYLAE. Pass of E. Greece lying between Mount Oeta and the Maliac Gulf, 9 mi. S.S.E. of Lamia, leading from Locris to Thessaly

    Hot Gates

Thermopylae of America, The. See: ALAMO, THE

Theseus, The Home of. See:

TROEZEN, GREECE

Thirst Land, The Great. See: SOUTH AFRICA

Thirteen Communes, The. Locality in province of Verona, Italy, near Badia

This Hemisphere, The Switzerland of. See: PANAMA

Thistle, The Land of the. See: SCOTLAND

Thousand Columns. See: Temple of a Thousand Columns, The

Thousand Islands, The Hundred. See: LACCADIVE ISLANDS

Thousand Isles, The River of a. See: WISCONSIN RIVER

Thousand Lakes, The Lake of the. See: SAIMAA, LAKE

Thousand Lakes, The Land of a. See: FINLAND

Three Cities, The. Collective appellation of cities of Cospicua, Senglea and Vittoriosa, all located on Valletta Harbor, island of Malta, in Mediterranean Sea S. of Sicily

Three Conference, The Place of the Big. See: YALTA, RUSSIA

Three C's Highway, The. See: C.C.C. HIGHWAY, THE

Three Kings, The City of the. See: COLOGNE, GERMANY

Three Peaks, The Mountain of. See: TITANO, MOUNT

Throat, Devil's. See: CROMER BAY

Throne of Jamsheed, The.
  See: PERSEPOLIS, PER-
  SIA

Throne of Solomon, The. See:
  TAKHT-SULAIMAN

THULE. Northernmost part
  of the habitable world as
  designated by the ancients:
  Norway, Iceland and/or
  Mainland, largest of the
  Shetland Islands

  Ultima Thule

THURGAU. Canton of N.E.
  Switzerland

  The Garden of Helvetia

TIBER RIVER. R. of central
  Italy flowing from Tuscan
  Apennines through city of
  Rome to the Tyrrhenian
  Sea

  Father Tiber

TIBERIAS, PALESTINE. An-
  cient town sit. on W. shore
  of Sea of Galilee, built by
  Herod Antipas in A.D. 21

  The Once-Royal City of
    Herod

Tiberias, The Sea of. See: GAL-
  ILEE, SEA OF

TIBET. Autonomous province
  of People's Republic of
  China sit. N. of Republic
  of India, Bhutan and Nepal

  The Forbidden Land

Tibet, Little. See: BALTIS-
  TAN, INDIA

Tides, The Bay of. See:
  FUNDY, BAY OF

TIEN-SHAN. Mountain sys-

tem of central Asia, ex-
  tending from the Pamir to
  the N. of the Tarim de-
  pression in Turkestan

  The Celestial Mountains

TIERRA. See: MAS A TIERRA.

TIFLIS, GEORGIA. Capital
  city of Georgian S.S.R. sit.
  on Dura R. 280 mi. W.N.W.
  of Baku

  Warm Springs

Tiger City. See: BHAGAL-
  PUR, INDIA

Tiger of the Alps, The. See:
  MATTERHORN, THE

Tight Little Island, The. See:
  ENGLAND; TAIWAN (FOR-
  MOSA)

Tight Little Isle, The. See:
  ENGLAND

TIMBUKTU, AFRICA. Town
  of Mali, W. Africa, sit. on
  edge of Sahara Desert

  The City of the Storks
  The Port of the Sudan in
    the Sahara

Time, The Land of Frozen.
  See: ANTARCTICA

TIMGAD, ALGERIA. Ruined
  city in dept. of Constantine,
  Algeria, 17 mi. S.E. of
  Batna

  The Numidian Pompeii

Tin Can Island. See: NIUA-
  FOO

Tin Pan Alley. Area around
  42nd St., New York City;
  site of offices of many mu-
  sic publishers

TITANO, MOUNT. Mountain
of San Marino Republic,
Italian peninsula; 2437 ft.
high

The Mountain of Three
Peaks

Tobacco, The Home of. See:
LATAKIA, SYRIA

Tobacco Coast, The. See:
CHESAPEAKE BAY

Toe, The. Southernmost part
of Italian peninsula

Toe of the Italian Boot, The.
See: CALABRIA, ITALY

TOKAY, HUNGARY. Town of
Zemplén county, Hungary,
about 130 mi. N.E. of Budapest

The Town of Tokay Wines

TOKYO, JAPAN. Capital
city of Japan, sit. on E.
coast of Island of Honshu
at head of Bay of Tokyo,
18 mi. N.E. of Yokohama

The Eastern Capital
The Gate of the Inlet

TOLEDO, SPAIN. Capital
city of Spanish province
of same name, sit. on
Tagus R. 45 mi. S.S.W.
of Madrid

The City of Sword Blades
The Stronghold of Carpe-
tani

Tom Quad. Quadrangle of
Christ Church College,
Oxford, England

Tombs, The. Former pris-
on sit. in New York
City, built in 1838 and

demolished in 1897

Tomorrow, The New World of.
See: EQUATORIAL AFRICA

TONGA ISLANDS. Independent
Polynesian kingdom sit. in
Pacific Ocean 180 mi. S.E.
of Fiji

The Friendly Islands

TONLE SAP. Lake sit. in W.
Cambodia, S.W. Indochina;
about 87 mi. long

The Great Lake

TOOWOOMBA, AUSTRALIA.
City of S.E. Queensland,
Australia, 65 mi. W. of
Brisbane

The Garden City of Queensland

Top of New England, The. See:
WHITE MOUNTAINS

Top of the World, The. See:
EVEREST, MOUNT; NORTH
POLE

TOR. See: MAM TOR

Tornado Alley. Broad belt in
continental United States
stretching from Texas to
Michigan; area where most
tornadoes of the country
occur

TORONTO, CANADA. Capital
city of province of Ontario,
Canada, sit. on N. shore
of Lake Ontario, 313 mi.
W.S.W. of Montreal

A City of Churches
A City of Homes
A Place of Meeting

TORRES STRAIT. Strait be-

tween island of New Guinea
and mainland of Australia
connecting Arafura Sea with
Coral Sea

The Straits

Tortoise, The Home of the.
See: GALAPAGOS ISLANDS

TOURAINE.   Historical region
of N.W. central France S.
of Le Maine

The Garden of France

Tourists, The Metropolis of.
See: LUCERNE, SWITZER-
LAND

TOURS, FRANCE.   Capital city
of the dept. of Indre-et-
Loire, France, sit. on left
bank of Loire R., 129 mi.
S.W. of Paris

The City of Silk
The City of the Turones

Tower.   See: Leaning Tower,
The

Tower, The.   See: TOWER
OF LONDON

Tower, The Bloody.   See:
TOWER OF LONDON

Tower, The City of the Leaning.   See: PISA, ITALY

TOWER OF LONDON.   Struc-
ture in London, England,
dating from reign of William
the Conqueror; long a prison for political offenders

The Bloody Tower
The Tower

Towers, The City of a Hundred.   See: PAVIA, IT-
ALY

TOWN.   See: AGAR-TOWN,
ENGLAND; SIMONS-TOWN,
SOUTH AFRICA

Town, George.   See: PENANG,
MALAYSIA

Town, The Maiden.   See:
EDINBURGH, SCOTLAND

Town, The New.   See: CARTH-
AGE, NORTH AFRICA

Town in South Africa, The
Most English.   See: GRA-
HAMSTOWN, SOUTH AF-
RICA

Town of Aristotle, The.   See:
STAGIRA, GREECE

Town of Banbury Tarts, The.
See: BANBURY, ENGLAND

Town of Brandy, The.   See:
COGNAC, FRANCE

Town of Cheese, The.   See:
ROQUEFORT-SUR-SOULZON,
FRANCE

Town of Cunobelin, The Royal.
See: COLCHESTER, ENG-
LAND

Town of Diamonds, The.   See:
KIMBERLEY, SOUTH AF-
RICA

Town of Floating Gardens, The.
See: XOCHIMILCO, MEX-
ICO

Town of Godiva's Ride, The.
See: COVENTRY, ENGLAND

Town of Hamlet, The.   See:
HELSINGÖR (ELSINORE),
DENMARK

Town of Horus, The.   See:
DAMANHUR, UNITED ARAB
REPUBLIC

Town of Humble Submission,
The. See: CANOSSA,
ITALY

Town of Many Marriages, The.
See: GRETNA GREEN,
SCOTLAND

Town of the Barisal Guns, The.
See: BARISAL, PAKISTAN

Town of the Beer Hall Putsch,
The. See: MUNICH, BA-
VARIA

Town of the Cudgel, The. See:
SHILLELAGH, IRELAND

Town of the Ford of the Hurdles, The. See: DUBLIN,
IRELAND

Town of the Kentishmen, The.
See: CANTERBURY, ENG-
LAND

Town of the Orpington Fowl,
The. See: ORPINGTON,
ENGLAND

Town of the Pied Piper, The.
See: HAMELN

Town of Tokay Wines, The.
See: TOKAY, HUNGARY

Town of Twelve Martyrs, The.
See: SCILLIUM, AFRICA

Town of Watchmakers, The.
See: FERNEY-VOLTAIRE,
FRANCE

Towns. See: Cautionary
Towns, The; Luther Towns,
The

TRACE. See: NATCHEZ
TRACE

Trade, The Sixteenth-Century
Emporium of South Amer-
ican. See: PORTO BEL-
LO, PANAMA

TRAIL. See: CHISHOLM
TRAIL

Trail, Boone's. See: WILDER-
NESS ROAD, THE

Trail, John's. See: CHIS-
HOLM TRAIL

Transport River, The. See:
GRAND CANAL

TRANSYLVANIA. Region of
N.W. and central Romania
sit. in Carpathian Mountains

The Gold Mine of Europe
The Home of the Vampire

Treason House. House be-
tween Stony Point and Haver-
straw, N.Y., used by Bene-
dict Arnold

Treasury of Peru, The. See:
ANDES MOUNTAINS

Treasury of Pharaoh, The.
See: EL KHAZNA

Treasury of Sciences, The.
See: BUKHARA, ASIA

Treaty, The City of the Violated. See: LIMERICK,
IRELAND

Treaty, The Stone of the Brok-
en. See: LIMERICK, IRE-
LAND

Trees. See: Mile of Christmas Trees, The

Trent, The Seat of the Council
of. See: TRENTO, ITALY

TRENTO, ITALY. Commune
of Trento province, N.E.
Italy, sit. on Adige R.,
106 mi. E.N.E. of Milan

The Seat of the Council of
Trent

Triangle. See: Bermuda Triangle, The; Golden Triangle, The

Triangle, The. Name applied to approx. 50,000 sq. mi. of E. Burma formed by N. Shan and S. Shan states and Karenna district

Tribes, The City of the. See: GALWAY, IRELAND

TRINITY COLLEGE. Educational institution sit. in Dublin, Ireland

The Silent Sister

TRIPOLI, LIBYA. Former Barbary state of N. Africa; now a part of Libya

The Barbary Pirates' Stronghold

TROEZEN, GREECE. Town of S.E. ancient Argolis, E. Peloponnesus, S. Greece

The Home of Theseus

TRONDHEIM, NORWAY. Capital and seaport of Sör-Trondelag county, Norway, sit. on S. side of Trondheim Fjord 250 mi. N. of Oslo

The City Where Kings Are Crowned

Trough, The Java. See: WHARTON DEEP

Trujillo, Ciudad. See: SANTO DOMINGO, HISPANIOLA

TRUK. Island group in central Caroline Islands, W. Pacific Ocean, 1500 mi. W. of Tarawa in Gilbert Islands

The Gettysburg of the Pacific

Tsar's Village, The. See: PUSHKIN, RUSSIA

TUCUMAN. Province of N. Argentina, federal republic of S. Central and S. South America

The Garden of the Argentine

TULA, RUSSIA. Capital of Tula region, U.S.S.R., 110 mi. S. of Moscow

The Birmingham of Russia

Tulip, The Land of the. See: NETHERLANDS, THE (HOLLAND)

TULLE, FRANCE. City of Corrèze dept., S. central France, 47 mi. S.S.E. of Limoges

The City of the Black Death

TUNGKWAN, CHINA. Town and fortress of E. Shensi, N. central China, sit. on Whang Ho R.

The Gateway to Shensi

TUNNEL. See: MOFFAT TUNNEL; SUTRO TUNNEL

Tunnel, The Great Gold. See: SUTRO TUNNEL

TURKEY. Republic sit. partly in S.E. Europe and partly in Asia Minor

The Sick Man
The Sick Man of Europe
The Sick Man of the East

Turkish Carpets, The Home of. See: USAK, TURKEY

Turn, The Lower. See:
VUELTA ABAJO

Turones, The City of the.
See: TOURS, FRANCE

Tusk, The Elephant. See:
INDIANOLA PEAK

Twelve Apostles, The. See:
APOSTLE ISLANDS

Twelve Islands. See: DO-
DECANESE ISLANDS

Twelve Martyrs, The Town of.
See: SCILLIUM, AFRICA

Two Nations, The Playground
of. See: GREAT SMOKY
MOUNTAINS

Two Rivers, The Country Between. See: MESOPO-
TAMIA

TYBURN HILL, ENGLAND.
Former place of execution
in London sit. on tributary
of Thames R. near the
present Hyde Park

The Hill of the Hangings

Tyler's Bailiwick, Wat. See:
DARTFORD, ENGLAND

TYRE, LEBANON. Town of
S. Lebanon sit. on Medi-
terranean Sea 50 mi. S. of
Beirut

The Biblical City
The Queen of the Sea

-U-

U.S.A. See: UNITED STATES
OF AMERICA

Ultima Thule. See: THULE

UMURBROGOL. Mountain on
Peleliu Island, Palau Islands,
W. Pacific Ocean

Bloody Nose Ridge

Under, Down. See: AUS-
TRALIA

Under the Ice, The City. See:
CAMP CENTURY, GREEN-
LAND

UNITED STATES OF AMERICA.
Federal republic of America
comprising 50 states and
District of Columbia

America
The Garden of the World
The Land of the Free and
the Home of the Brave
The Land of the Stars and
Stripes
The Mainland
The States
U.S.A.
The United States

United States, The Capital of
the Negro Population of the.
See: HARLEM

Universe. See: Hub of the
Universe, The

USAK, TURKEY. Manufactur-
ing town of Kütahya vilayet,
W. Turkey, 55 mi. W. of
Afyon Karahisar

The Home of Turkish Car-
pets

Useless Bay. See: INUTIL
BAY

USU. See: SHABARAKH USU

-V-

Vale of Kashmir, The. See:

KASHMIR (CASHMERE)

VALLEY. See: IMPERIAL
VALLEY; SAN JOAQUIN
VALLEY; TENNESSEE VAL-
LEY; WEI RIVER VALLEY

Valley. See: Golden Valley,
The; Magic Valley, The

Valley, The Happy. See:
KASHMIR (CASHMERE);
TENNESSEE VALLEY

Valley of Jehoshaphat, The.
See: KIDRON WADI

Valley of Norway, The East.
See: OSTERDAL

Valley of Stones, The. See:
GUADALAJARA, SPAIN

Valley of the Kings, The.
Rocky ravine near Thebes,
Egypt; burial place of an-
cient Egyptian pharaohs

Valley of Wonders, The. See:
YELLOWSTONE NATIONAL
PARK

Vampire, The Home of the.
See: TRANSYLVANIA

VATICAN CITY. Independent
papal state within commune
of Rome, Italy, sit. on
right bank of Tiber R.

The Home of His Holiness
The Home of the Pope

Vatican of Buddhism, The.
See: MANDALAY, BURMA

Vaunted Island, The. See:
SINGAPORE ISLAND

Veda, The Land of the. See:
INDIA

VENICE, ITALY. Seaport of
Venezia province, N.E.
Italy, on 118 islands in
Lagoon of Venice, 162 mi.
E. of Milan

The Bride of the Adriatic
The Bride of the Sea
The City of the Lagoons
The Mistress of the Adri-
atic
The Queen of the Adriatic
The Silent City

Venice, Little. See: AMIENS,
FRANCE; ARENDAL, NOR-
WAY

Venice, The Provençal. See:
MARTIGUES, FRANCE

Venice of America, The. See:
RECIFE, BRAZIL

Venice of China, The. See:
WUHSIEN, CHINA

Venice of France, The. See:
AMIENS, FRANCE

Venice of Japan, The. See:
OSAKA, JAPAN

Venice of the East, The. See:
BANGKOK, THAILAND;
WUHSIEN, CHINA

Venice of the West, The. See:
GLASGOW, SCOTLAND

Venus, The City of. See:
MELOS, GREECE

Venus's Court. See: HÖRSEL
BERGE, GERMANY

VERACRUZ, MEXICO. Town
and port in state of Veracruz, on E. coast of Mexico

The City of the Dead

VERDE, CAPE. Promontory
on coast of W. Africa sit.

between the Senegal and
Gambia Rivers

Africa's Farthest West

Vermillion Sea, The. See:
CALIFORNIA, GULF OF

Vernon. See: Mount Vernon

Versailles of Naples, The.
Cathedral palace sit. at
Caserta, Napoli province,
Campania, S. Italy; de-
signed by Vanvitelli and
begun in 1752

Versailles of Poland, The.
Palace of Counts of Bran-
ski sit. in town of Biały-
stok, Poland

Via Dolorosa (The Way of
Pain). Road from Mount
of Olives to Golgotha,
which Christ traveled on
way to Crucifixion

Victorious City, The. See:
CAIRO, EGYPT

Victory, The City of. See:
CAIRO, EGYPT

Victory, The City of Pyrrhic.
See: HERACLEA, ITALY

VIENNA, AUSTRIA. Capital
and largest city of Austria
sit. on both banks of Danube R. 380 mi. S.S.E. of
Berlin

The City of a Million
Dreams
The Paris of Eastern
Europe

VIET NAM. Independent re-
public of Asia comprising
N. Annam and Tonkin

The Distant South

The Divided Dragon
The Land of Many Dragons
The Southern Land

Villa of Tears, The. House
in suburbs of Coimbra,
Portugal; scene of death
of Inez de Castro

VILLAGE. See: GREENWICH
VILLAGE

Village, The Children's. See:
PUSHKIN, RUSSIA

Village, The Tsar's. See:
PUSHKIN, RUSSIA

Village of the Marsh, The.
See: BRUSSELS, BELGIUM

VINCENT. See: SAINT VIN-
CENT, CAPE

Violated Treaty, The City of
the. See: LIMERICK, IRE-
LAND

Violet Crown, The City of the.
See: ATHENS, GREECE

Violet-Crowned Athens, Ancient.
See: ATHENS, GREECE

Violet-Crowned City, The.
See: ATHENS, GREECE

Violin City, The. See: CRE-
MONA, ITALY

VLADIMIR, U.S.S.R. City of
Vladimir region, Soviet Russia, Europe, sit. on Klyaz-
ma R., 110 mi. E. of Mos-
cow

The Coronation City

VOGELKOP. Peninsula com-
prising extension of Netherlands New Guinea sit. in
S.W. Pacific Ocean N. of
McCluer Gulf

The Bird's Head

VOLGA RIVER. River of So-
viet Russia in Europe flowing from Valdai Hills in N.
Kalinin region to Caspian
Sea; 2325 mi. long

The Great Volga

VOLTAIRE. See: FERNEY-
VOLTAIRE, FRANCE

VUELTA ABAJO. Section of
island of Cuba W. of Mer-
idian of Havana, largely in
Pinar del Rio province

The Lower Turn

-W-

WADI. See: KIDRON WADI

WAIKIKI BEACH. Beach on
Pacific Ocean at Honolulu,
Hawaii; tourist attraction

The Long Branch of Hon-
olulu

WAIMEA CANYON. Canyon
sit. on Kauai Island, Ha-
waii; 3,000 ft. deep

The Grand Canyon of the
Pacific

Walcheren, The Netherlands.
See: Cautionary Towns,
The

WALES. Principality forming
peninsula on W. of island
of Great Britain

The Land of the Leek

Wales, The Lourdes of. See:
HOLYWELL, WALES

Wales, One of the Seven Won-

ders of. See: SAINT
GILES'S CHURCH

Walk. See: Birdcage Walk;
Long Walk, The

WALL. See: BERLIN WALL;
GREAT WALL, THE

Wall. See: Devil's Wall, The

Wall, The Anti-Fascist Protection. See: BERLIN
WALL

Wall, The Chinese. See:
GREAT WALL, THE

Wall of China, The Great.
See: GREAT WALL, THE

Wall of Infamy, The. See:
BERLIN WALL

WALL STREET, NEW YORK
CITY. Street of New York
City; center of financial dis-
trict

The Street of Extremes

Wall Street of Boston. See:
STATE STREET, BOSTON,
MASS.

Wall Street of Chicago, The.
See: LA SALLE STREET,
CHICAGO, ILL.

Wall Street of Detroit, The.
See: GRISWOLD STREET,
DETROIT, MICH.

Wall Street of the West, The.
See: MONTGOMERY STREET,
SAN FRANCISCO, CALIF.

Wallenstein's Death, The City
of. See: CHEB, CZECHO-
SLOVAKIA

WAN-SHOU-SHAN. Park of N.
Hopeh province, sit. in N.E.
China, 8 mi. N.W. of Peking

The Mountain of Ten Thousand Ancients

WANCHUAN, MONGOLIA.
Capital city of Chahar province, E. Inner Mongolia, N. China, sit. near Nankow Pass of the Great Wall of China

The Gate to Mongolia

Wapping of Denmark, The.
See: HELSINGÖR (ELSINORE), DENMARK

War Lord, The Seat of the Mukden. See: MUKDEN, MANCHURIA

WARE, ENGLAND. Market town of Hertfordshire, S.E. England, sit. on Lea R., 23 mi. N. of London

The Home of the Great Bed

Warm Springs. See: TIFLIS, GEORGIA

WARWICKSHIRE. County of central England

The Heart of England

WASHINGTON SQUARE.
Small park sit. on Dearborn and Walton Place, near N. side of Chicago, Ill.

Bughouse Square

Waste, The Sandy. See: GOBI DESERT

Wat Tyler's Bailiwick. See: DARTFORD, ENGLAND

Watchmakers, The Town of. See: FERNEY-VOLTAIRE, FRANCE

Watering Places, The Queen

of English. See: SCARBOROUGH, ENGLAND

Waterloo, Germany's. See: BASTOGNE, BELGIUM

Waterloo, Hitler's. See: BASTOGNE, BELGIUM

Waterloo, Montcalm's. See: PLAINS OF ABRAHAM

Waters, The Father of. See: IRRAWADDY RIVER; MISSISSIPPI RIVER

Waters, The King of. See: AMAZON RIVER; MISSISSIPPI RIVER

Waters, The Place of the Muddy. See: SHABARAKH USU

WATLINGS ISLAND. See: SAN SALVADOR (WATLINGS ISLAND)

WAY. See: APPIAN WAY

WAY (Way). See also: WEI

Way, The Great White. See: BROADWAY, NEW YORK CITY

Way of Pain, The. See: Via Dolorosa (The Way of Pain)

Wayward Sisters, The. Appellation of seceding southern states during American Civil War

Weddings. See: House of Golden Weddings, The

WEENEN, SOUTH AFRICA.
Town of central Natal, S. Africa, 85 mi. N.W. of Durban

The Place of Weeping

Weeping, The Place of. See:

WEENEN, SOUTH AFRICA

WEI. See also: WAY; Way

WEI RIVER VALLEY. Valley of N. central China

The Cradle of Chinese Civilization

WEIMAR, GERMANY. City of Thuringia, Germany, 13 mi. E. of Erfurt

The German Athens

WELFARE ISLAND. Island in E. River, N.Y.; former penal institution

Blackwell's Island
The Island

Well, The Isle of the Leper's. See: STRONSAY

Wells of Itza, The Mouth of the. See: CHICHEN ITZA

West, Africa's Farthest. See: VERDE, CAPE

West, The Athens of the. See: CORDOVA, SPAIN

West, Australia's Farthest. See: STEEP POINT

West, The City of the. See: GLASGOW, SCOTLAND

West, The Emerald Island of the. See: MONTSERRAT

West, The Empire of the. See: HOLY ROMAN EMPIRE

West, The Far. See: MOROCCO

West, The Hudson River of the. See: MISSISSIPPI RIVER

West, Norway's Farthest. See: STEINSÖY, NORWAY

West, The Venice of the. See: GLASGOW, SCOTLAND

West, The Wall Street of the. See: MONTGOMERY STREET, SAN FRANCISCO, CALIF.

WEST END. Aristocratic part of London, England, including S. Hyde Park and Mayfair

The London of Fashion and Pleasure

West Point of the Air, The. See: RANDOLPH FIELD, TEXAS

West River. See: SI-KIANG

West Side. W. part of borough of Manhattan, New York City

Western Lake, The. See: SI-HU

Western Ocean, The. See: ATLANTIC OCEAN

Western White House, The. See: SAN CLEMENTE, CALIFORNIA

Western White House, The Former. See: SAN CLEMENTE, CALIFORNIA

Westernmost Point of Africa, The. See: ALMADIES, CAPE

Westernmost Point of Europe, The. See: SAINT VINCENT, CAPE

Westminster Abbey. See: COLLEGIATE CHURCH OF ST. PETER

Wetbacks, The River of the.
See: RIO GRANDE

WHARTON DEEP. One of
deepest known parts of
Pacific Ocean (22,968 ft.)
lying off S. coast of island
of Java

The Java Trough

Wheels. See: Hell on Wheels

Where Kings Are Crowned,
The City. See: TROND-
HEIM, NORWAY

Where Past and Present Meet.
See: CAMBODIA

White Elephant, The Land of
the. See: THAILAND

WHITE HOUSE. Official residence of Presidents of the
United States sit. in Washington, D.C.

The Yankee Palace

White House, The Former
Western. See: SAN
CLEMENTE, CALIFORNIA

White House, The Western.
See: SAN CLEMENTE,
CALIFORNIA

White Mountain. See: MAUNA
KEA; SAFED KOH

WHITE MOUNTAINS. Moun-
tains of Appalachian range
sit. in S. Maine and N.
New Hampshire

The Crystal Hills
The Top of New England

White River, The. See:
PEI-HO

White-Stoned, The. See:
MOSCOW, RUSSIA

White Way, The Great. See:
BROADWAY, NEW YORK
CITY

White Wines, The Home of.
See: CHABLIS, FRANCE

White Woman, The. See:
IXTACIHUATL

Whit's Palace. See: NEW-
GATE PRISON

WIGHT, ISLE OF. Island and
administrative county of
England, sit. in English
Channel

The Garden of England

Wilderness, The Flow of.
See: GALAPAGOS ISLANDS

WILDERNESS ROAD, THE.
Road from E. Virginia
through Cumberland Gap
and central Kentucky to
Ohio R.

Boone's Trail

William Tell, The Home of.
See: ALTDORF, SWITZER-
LAND

WILSON'S PROMONTORY.
Promontory on central S.
coast of Australia S.E. of
Melbourne

Australia's Farthest South

WILTSHIRE. County of S. England

The Home of the Moonrakers

WINDERMERE. Lake $10\frac{1}{2}$ mi.
long sit. in Lake District,
N.W. England; largest lake
in England

The Queen of the Lakes

Window into Europe, A. See:
LENINGRAD, U.S.S.R.

Windsor of Denmark, The.
Appellation of Castle of
Kronborg, sit. at Elsinore,
seaport of N. Sjaelland Island, Denmark

Wine, The Home of Neckar.
See: ESSLINGEN, WEST
GERMANY

Wine, The Home of Port.
See: OPORTO, PORTUGAL

Wine Land. See: OENOTRIA,
ITALY

Wines, The Home of. See:
BORDEAUX, FRANCE

Wines, The Home of Madeira.
See: MADEIRA

Wines, The Home of White.
See: CHABLIS, FRANCE

Wines, The Town of Tokay.
See: TOKAY, HUNGARY

Winter Garden, America's
Great. See: IMPERIAL
VALLEY

Wipers. See: IEPER
(YPRES), BELGIUM

WISCONSIN RIVER. River
of S.W. Wisconsin flowing
from Lac Vieux Desert to
Mississippi R.; about 430
mi. long

The River of a Thousand
Isles

Wisdom, The Land of. See:
NORMANDY

Wise Men, The Home of the.
See: GOTHAM, ENG-
LAND

Within a Desert, The Desert.
See: TANEZROUFT

WITTENBERG, GERMANY.
City of Halle district, E.
Germany, sit. on Elbe R.
55 mi. S.W. of Berlin

The Cradle of the Reformation
The Focal Point of the Reformation
The Home of the Reforma-
tion

WITWATERSRAND. Geograph-
ical region of Transvaal
province, Republic of S.
Africa

The Rand

Wolfland. See: IRELAND
(EIRE)

Woman, The Fat. See: IXTA-
CIHUATL

Woman, The White. See: IX-
TACIHUATL

Wonder of the World, The
Eighth. See: BROOKLYN
BRIDGE

Wonderland of America, The.
See: YELLOWSTONE NA-
TIONAL PARK

Wonders, The Valley of. See:
YELLOWSTONE NATIONAL
PARK

Wonders of the Ancient World.
See: Seven Wonders of the
Ancient World, The

Wonders of the World. See:
Seven Wonders of the World,
The

Wonders of Wales, One of the
Seven. See: SAINT GILES'S
CHURCH

WORCESTER, ENGLAND.
Borough and administra-
tive center of Worcester-
shire, England, sit. on
Severn R., 100 mi. N.W.
of London

The Home of Worcester-
shire Sauce

WORCESTERSHIRE, ENGLAND.
County of W. central Eng-
land

The Garden of England

Worcestershire Sauce, The
Home of. See: WORCES-
TER, ENGLAND

World. See: New World,
The; Old World, The;
Seven Wonders of the An-
cient World, The; Seven
Wonders of the World, The

World, The Banker to Half the.
See: HONG KONG, CHINA

World, The Bottom of the.
See: ANTARCTICA; SOUTH
POLE

World, The Center of the. See:
LAOS

World, The Dardanelles of the
New. See: DETROIT RIV-
ER

World, The Diamond Capital
of the. See: ANTWERP,
BELGIUM

World, The Eighth Wonder of
the. See: BROOKLYN
BRIDGE

World, The Garden of the.
See: UNITED STATES
OF AMERICA

World, The Gibraltar of the

New. See: DIAMOND,
CAPE

World, The Goddess-Mother
of the. See: EVEREST,
MOUNT

World, The Granary of the
Roman. See: EGYPT

World, The Hub of the. See:
GIBRALTAR

World, The Mistress of the.
See: ROME, ITALY

World, The Oil Capital of the.
See: DHAHRAN, SAUDI
ARABIA

World, The Paradise of the.
See: CONGO, REPUBLIC
OF THE

World, The Paris of the Ancient. See: CORINTH,
GREECE

World, The Queen of the. See:
MEROË

World, The Rice Bowl of the.
See: BANGKOK, THAILAND

World, The Roof of the. See:
PAMIR

World, The Sea of the New.
See: CARIBBEAN SEA

World, The Smallest Republic
in the. See: SAN MARINO

World, The Summit of the.
See: EVEREST, MOUNT;
YELLOWSTONE NATIONAL
PARK

World, The Top of the. See:
EVEREST, MOUNT; NORTH
POLE

World Ditch, The. See: SUEZ
CANAL

World of Tomorrow, The
New. See: EQUATORIAL
AFRICA

World's Diamond Center, The.
See: KIMBERLEY, SOUTH
AFRICA

WORMS, GERMANY. City and
state of Rhineland Palatinate,
W. Germany, sit. on W.
bank of Rhine R., 10 mi.
N.N.W. of Mannheim

The Mother of Diets

Wuhan. See: Han Cities, The

WUHSIEN, CHINA. City and
port of Kiangsu province,
China, sit. on Grand Canal,
55 mi. W. of Shanghai

The Venice of China
The Venice of the East

-X-

XOCHIMILCO, MEXICO.
Town of central Mexico,
10 mi. S. of Mexico City
sit. on W. shore of Lake
Xochimilco

The Town of Floating Gar-
dens

-Y-

YALTA, RUSSIA. Town of
Crimean region, Ukrainian
Soviet Russia, Europe, sit.
on Black Sea 30 mi. E. of
Sevastopol

The Center of the Soviet
Riviera
The Place of the Big Three
Conference

YALU RIVER. River of Asia
forming part of boundary
between Korea and Manchuria;
about 300 mi. long

The Dividing Line

Yang Kingdom River, The.
See: YANGTZE KIANG
RIVER

YANGTZE KIANG RIVER.
Principal R. of China, flow-
ing from E. Kunlun Mts. in
S.W. Tsinghai to E. China
Sea near Shanghai; about
3200 mi. long

The Girdle of China
The Yang Kingdom River

Yankee Palace, The. See:
WHITE HOUSE

YAP. Island of W. Caroline
Islands sit. in W. Pacific
Ocean 500 mi. S.W. of
Guam

The Island of Stone Money

YAWATA, JAPAN. Seaport
city of Fukuoka province,
Japan, about 13 mi. S.W.
of Moji

The Pittsburgh of Japan

Yellow River, The. See:
HWANG HO

YELLOWSTONE NATIONAL
PARK. U.S. national park
sit. in N.W. Wyoming, S.
Montana and E. Idaho

The Summit of the World
The Valley of Wonders
The Wonderland of America

Yen, The Land of the Rising.
See: JAPAN

YERBA BUENA. Island in

San Francisco Bay, Calif.

Goat Island

Youth, The Island of the Fountain of. See: BIMINI

YPRES. See: IEPER (YPRES),
BELGIUM

-Z-

Zaleucus, The City of. See:
LOCRI, ITALY

ZANTE. Ionian island off
N.W. coast of Pelopon-
nesus, 8 mi. S. of Cepha-
lonia

The Flower of the Levant

Zeus, The City of. See:
THEBES, EGYPT

Zion. See: TEMPLE HILL

ZION, PALESTINE. Height in
N.E. part of city of Jerusalem, Palestine, originally
the Jebusite stronghold; cap-
tured by David

The City of David

ZOUTPANSBERG. Mountain
range in N. Transvaal, N.E.
Union of S. Africa

Salt Pan Hill

ZURICH, SWITZERLAND. Capital city of canton of same
name, Switzerland, 60 mi.
N.E. of Bern

The Athens of Switzerland